Vorwort zur Patentsammlung Puppenhäuser

Seit dem frühen Mittelalter begleiten Musterbücher und Musterblätter große Meister und ihre Familien. In ihnen wurde das über Generationen gesammelte Wissen bewahrt und an die Nachfolger und Erben weitergegeben. Patentschriften sind moderne Musterbücher für Modellbauer. Patente sind die jeweils weltweit erste Dokumentation einer neuen Idee und belegen auf besondere Art das Fortschreiten der Technik und Ihre Geschichte. Aus der riesigen internationalen Sammlung technischer Lösungen mit über 200 einzigartigen innovativen Ideen zum Thema Puppenhaus wurden die vorliegenden repräsentativen Patentschriften ausgewählt und zu einem besonderen Musterbuch zusammengeführt. Jede Schrift enthält eine umfassende technische Beschreibung der Erfindung und detailgenaue Zeichnungen. Konzept, Konstruktion und Funktion lassen sich anhand der Darstellung in allen Details nachvollziehen.

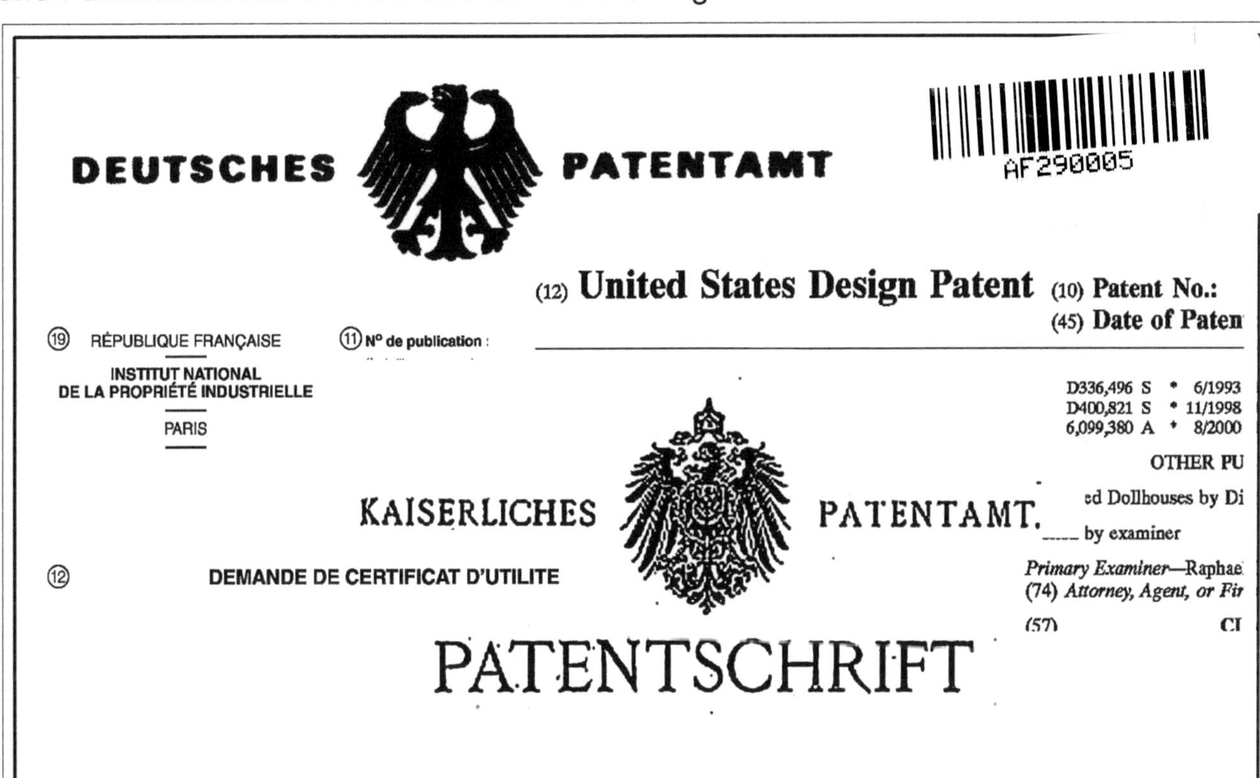

Ungefähr 90 % des weltweiten technischen Wissens ist in Patentschriften veröffentlicht. Von diesen Patenten sind jedoch nur ca. 10% in Kraft. Bei über 90 % der registrierten Erfindungen ist der Rechtsschutz inzwischen verfallen und die Nutzung kostenlos möglich. Alle Puppenhaus-Patente, die in dieser Sammlung zusammengetragen wurden, sind mit Erscheinen dieses Buches frei von Patentrechten. Die technischen Problemlösungen sind somit frei von Schutz-Rechten und können jederzeit privat genutzt werden. Die gewerbliche Verwendung kann trotzdem weiterhin verwandten Schutzrechten wie z.B.: dem unlauteren Wettbewerb UWG unterliegen und somit eingeschränkt sein.

Zeichnen Sie Ihre eigenen Bilder, entwerfen Sie Ihr eigenes Werk,
und werden Sie selbst ein Schöpfer!
Denn die IDEE ist frei und unterliegt keinem Schutzrecht!

St. Helena, 07. August 2014 K. Winter

Bibliographische Information der Deutschen Nationalbibliothek
Die Deutsche Nationalbibliothek verzeichnet diese Publikation in der Deutschen Nationalbibliografie;
detaillierte bibliografische Daten sind im Internet über http://dnb.d-nb.de abrufbar.

Reihe: Vorlage - Konzept - Konstruktion
Band 1, 1. Auflage 2014

Lektorat sowie Entwurf und Gestaltung,
Arrangement der Texte, Bilder und Grafiken
durch ATELIER-KALAI

Herstellung und Verlag
Books on Demand GmbH, Norderstedt
ISBN 978-3-7357-9413-0

INHALTSANGABE
Patentsammlung Puppenhäuser I
Design und Technik

Patentsammlung Puppenhäuser I • Design und Technik • Reihe: Vorlage - Konzept - Konstruktion • Band: 1
Herausgeber: www.atelier-kalai.de, Kerstin Winter • Hersteller / Verlag: Books on Demand GmbH Norderstedt

Bisher bei Atelier–Kalai erschienen:

• Kunstschmieden / Ironwork / Ferronnerie / Hierro Forjado I
Gitter und Geländer aus Schmiedeeisen
(sehen-wissen-gestalten Band 1)

• Musterbuch Renaissance Stickereien
Stickvorlagen für Blumen Tiere und Waldfrüchte
(sehen-wissen-gestalten Band 2, 2. Aufl)

• Musterbuch Blumen sticken
(sehen-wissen-gestalten Band 3)

• Chinesische Gartenhäuser und Pavillons
(sehen-wissen-gestalten Band 4)

• Patentsammlung Puppenhäuser I
(Vorlage-Konzept-Konstruktion Band 1)

• Patentsammlung Puppenhäuser II
(Vorlage-Konzept-Konstruktion Band 2)

Patentsammlung Puppenhäuser I • Design und Technik • Reihe: Vorlage - Konzept - Konstruktion • Band: 1
Herausgeber: www.atelier-kalai.de, Kerstin Winter • Hersteller / Verlag: Books on Demand GmbH Norderstedt

KAISERLICHES PATENTAMT.

PATENTSCHRIFT

— № 200316 —

KLASSE 77 f. GRUPPE 23.

MAX GANTNER in STUTTGART.

Puppenhaus aus zwei gelenkig miteinander verbundenen Teilen.

Patentiert im Deutschen Reiche vom 4. Juli 1907 ab.

Den Gegenstand der Erfindung bildet ein Puppenhaus, das aus zwei gelenkig miteinander verbundenen Teilen besteht.

Das Puppenhaus zeichnet sich dadurch aus, daß die gelenkig miteinander verbundenen Teile auf einem gemeinsamen Sockel angeordnet sind, auf dem sie sich um in Schlitzen verschiebbaren Zapfen so drehen, daß sie in geschlossenem Zustand die Außenfront eines vollständigen Hauses darstellen, während sie in geöffnetem oder auseinandergeklapptem Zustand die innere Einrichtung des Hauses zeigen.

Fig. 1 der Zeichnung zeigt ein Ausführungsbeispiel des Puppenhauses in geschlossenem Zustand, Fig. 2 einen wagerechten Schnitt dazu. Fig. 3 zeigt das Puppenhaus geöffnet und Fig. 4 den wagerechten Schnitt hiervon.

Das Puppenhaus besteht aus zwei gelenkig miteinander verbundenen Teilen a, b. Um die Teile a, b auseinanderzuklappen, d. h. aus der Lage Fig. 1 in die von Fig. 3 zu bringen, müssen die Teile a, b gedreht werden. Die Bewegung der beiden Teile a, b durch irgendwelche Kraft, beispielsweise ein Federtriebwerk, kann aber auch so vor sich gehen, daß die Teile a, b durch eine Zahnstange zunächst um ihre Breite nach vorwärts bewegt und dann um 90° nach auswärts gedreht werden.

Die dargestellte Bewegungsvorrichtung zum Drehen der beiden Teile besteht im wesentlichen aus einer auf geeignete Art angetriebenen Spindel k, die mit Rechts- und Linksgang versehen ist, um gleichzeitig eine Vor- und Rückwärtsbewegung zu erzielen. Die Spindel k läuft in einem Schieber i, der auf einer Platte f sitzt. Die Platte f trägt zwei Zapfen g, h, welche in die Teile a, b eingreifen und diesen als Scharnier dienen und sie mitnehmen. Ferner besitzt jeder der Teile a, b im Mittelpunkt seines Bodens einen Zapfen c, der sich in einem Schlitz d des Sockels e führt. Wird nun die Spindel gedreht, so müssen die Teile a, b eine Drehung um die in den feststehenden Schlitzen d des Sockels gleitenden Zapfen c ausführen.

Die Bewegung der Spindel erfolgt stets in einer Drehrichtung; wenn die Teile a, b in die geschlossene oder in die geöffnete Lage angekommen sind, wird das die Spindel bewegende Triebwerk auf beliebige Weise abgestellt.

PATENT-ANSPRÜCHE:

1. Puppenhaus aus zwei gelenkig miteinander verbundenen Teilen, dadurch gekennzeichnet, daß die beiden Teile (a, b) auf einem gemeinsamen Sockel (e) angeordnet sind, auf dem sie durch eine mechanische Vorrichtung um in Schlitzen verschiebbare Zapfen (c) so gedreht werden können, daß sie aus der geschlossenen, die Außenfront eines Hauses zeigenden Stellung in die geöffnete, die innere Einrichtung des Hauses zeigende Stellung gelangen, wobei in beiden Stellungen der Sockel symmetrisch zu beiden Teilen liegt.

2. Puppenhaus nach Anspruch 1, dadurch gekennzeichnet, daß die beiden Teile (a, b) durch eine in irgendeiner Weise angetriebene Spindel (k) und einen auf ihr laufenden Schieber (i) gedreht werden, der mit den auf einer Fläche (f) sitzenden Zapfen (g, h) in die beiden Teile (a, b) eingreift.

Hierzu 1 Blatt Zeichnungen.

Patentsammlung Puppenhäuser I • Design und Technik • Reihe: Vorlage - Konzept - Konstruktion • Band: 1
Herausgeber: www.atelier-kalai.de, Kerstin Winter • Hersteller / Verlag: Books on Demand GmbH Norderstedt

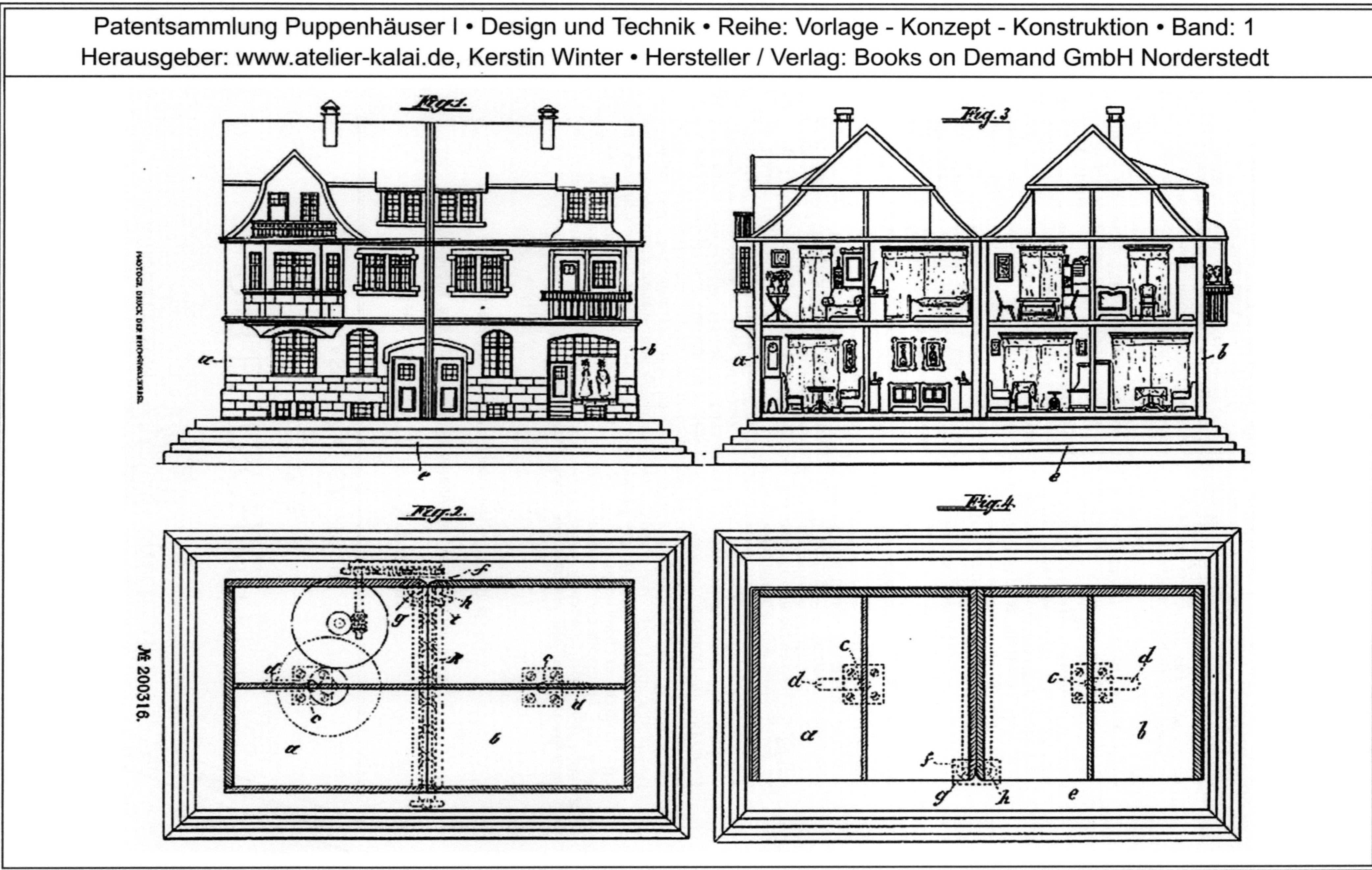

United States Patent Office.

SUSAN DIANA ROSA, OF EDINBURGH, SCOTLAND.

DOLL'S HOUSE.

SPECIFICATION forming part of Letters Patent No. 688,008, dated December 3, 1901.

Application filed April 4, 1900. Serial No. 11,434. (No model.)

To all whom it may concern:

Be it known that I, SUSAN DIANA ROSA, gentlewoman, a subject of the Queen of the United Kingdom of Great Britain and Ireland, and a resident of 10 Mayfield road, Edinburgh, Scotland, have invented new and useful Improvements in Dolls' Houses, of which the following is a specification.

This invention relates to improvements in dolls' houses; and it consists in so constructing the sides of the house that they can be opened while the front of the house remains intact and a fixture, enabling a child to play with the dolls in the various rooms and at the same time have the use of the street or front door, by which the dolls may be passed in and out, the front door remaining intact in perfect simulation of the front and entrance door of an actual house.

Figure 1 shows the one side of the house opened out. Fig. 2 shows the same side closed. Fig. 3 shows the side folded up compact against the side of the house and out of the way against the side of the house.

As shown in the said drawings, the sides are made in sections A, hinged together either by linen strips *a* or other hinges, so that they can be folded compact and tied together by a band or chain B, which embraces them, the free end of which can be attached to the hook *b*, the folded side being thus entirely out of the way. When the side is closed, as in Fig. 2, the various sections of the side may be held rigid by a cross-bar C, hinged or pivoted at *c* and provided at the other end with a notch or recess D, engaging with a pin *d*, as shown in Fig. 2.

I claim—

1. A doll's house constructed with an immovable front; several interior rooms separated by transverse partitions and hinged side walls adapted to be opened to give access to the interior rooms, as described.

2. A doll's house constructed with an immovable front; several interior rooms separated by partitions; hinged side walls made up of a number of vertical leaves A connected together by flexible strips *a* so that such side wall when opened may be folded, accordion-like, within a small space and bars D by which said hinged walls are rigidly secured when closed, as described.

3. A doll's house having a fixed front; several stories of interior rooms separated by transverse partitions; and hinged side walls each consisting of a number of vertical leaves A, A, extending from bottom to top, connected by flexible strips *a*, *a*, adapting the entire side of the house to be folded back in compact form out of the way so as to afford access to the several rooms of the different stories or to be rigidly secured in closed position, as shown and described.

4. In a doll's house, the combination of a fixed front having a front door; an interior structure of several stories consisting of separate rooms separated by transverse partitions; hinged and folding sides each consisting of a number of vertical leaves A, A, extending from bottom to top, connected by flexible strips *a*, *a*, adapting the entire side wall to be folded back into compact position out of the way and afford access to all the interior rooms; bands B for securing the hinged sides in folded position; and bars C by which the said hinged sides are rigidly secured in extended and closed position, all as shown and described.

SUSAN DIANA ROSA.

Witnesses:
THOS. J. CONNOLLY,
JOHN TAIT.

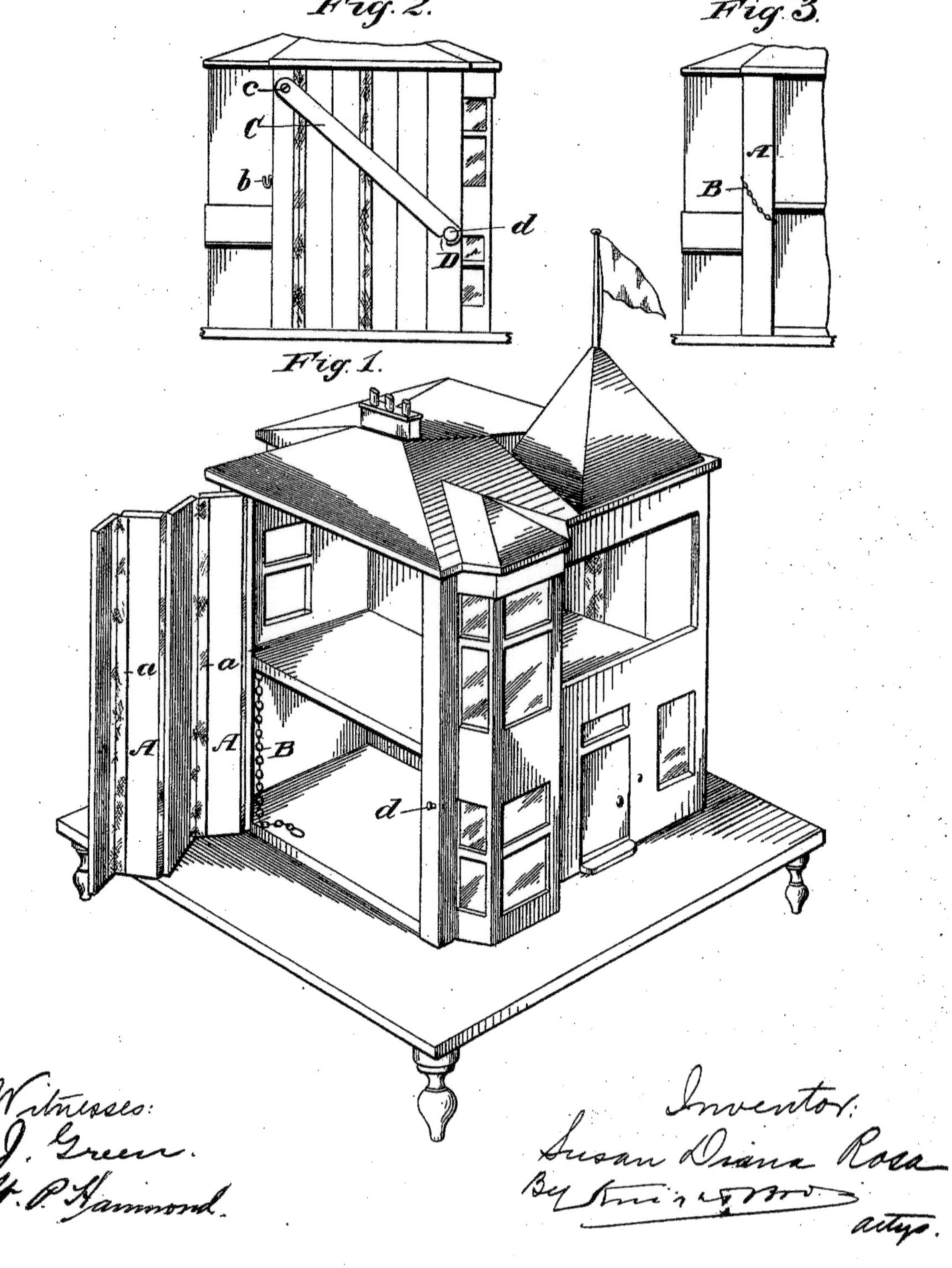

Witnesses:
J. Green.
W. P. Hammond.

Inventor:
Susan Diana Rosa
By
attys.

Patentsammlung Puppenhäuser I • Design und Technik • Reihe: Vorlage - Konzept - Konstruktion • Band: 1
Herausgeber: www.atelier-kalai.de, Kerstin Winter • Hersteller / Verlag: Books on Demand GmbH Norderstedt

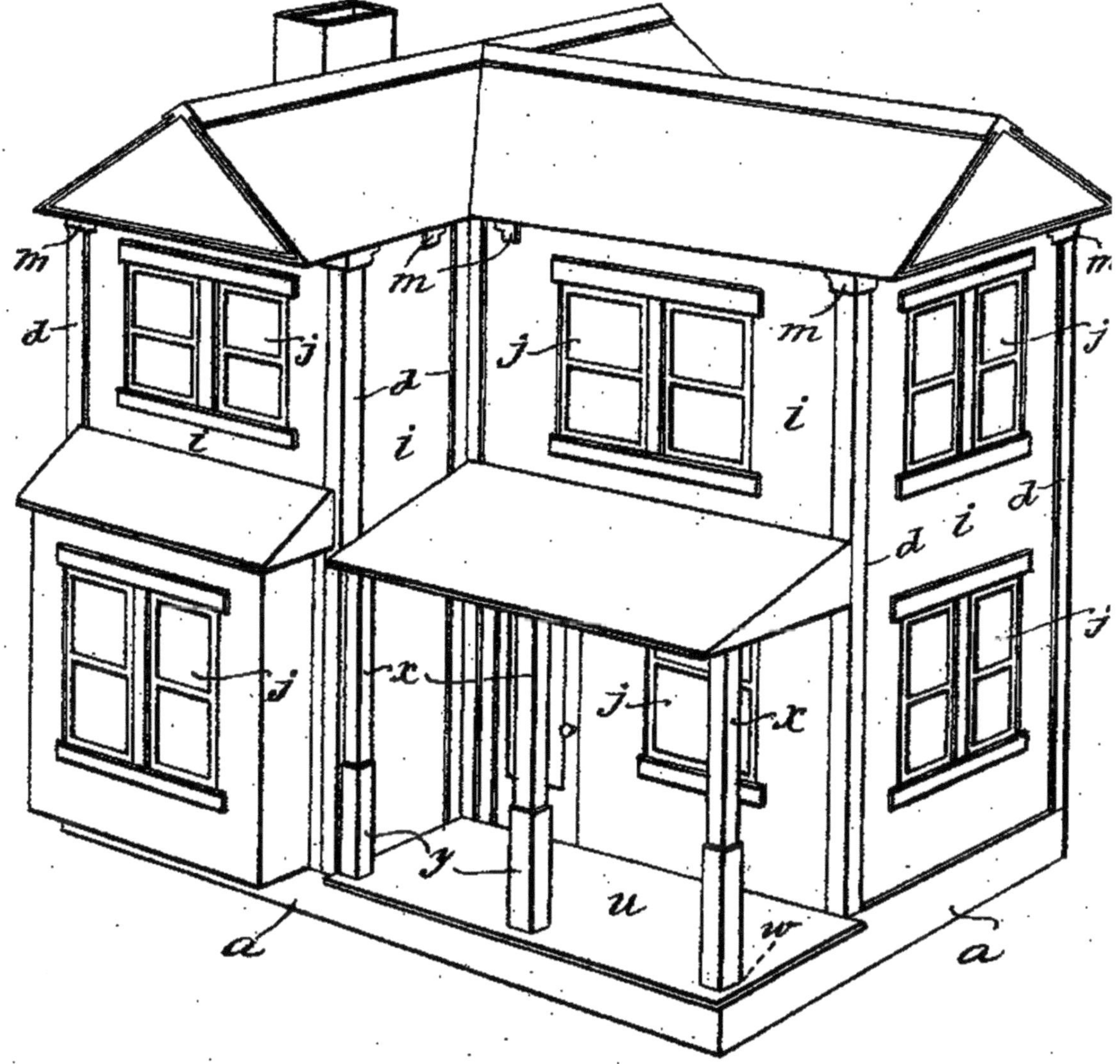

No. 739,669.
PATENTED SEPT. 22, 1903.
G. E. GRIMM.
KNOCKDOWN TOY HOUSE.
APPLICATION FILED JULY 26, 1902.
NO MODEL.
6 SHEETS—SHEET 1.
Fig. 1
Witnesses:
W. A. Schaefer
Inventor.
Gerhardt E. Grimm.
By his Attorney
Chas. A. Cutter

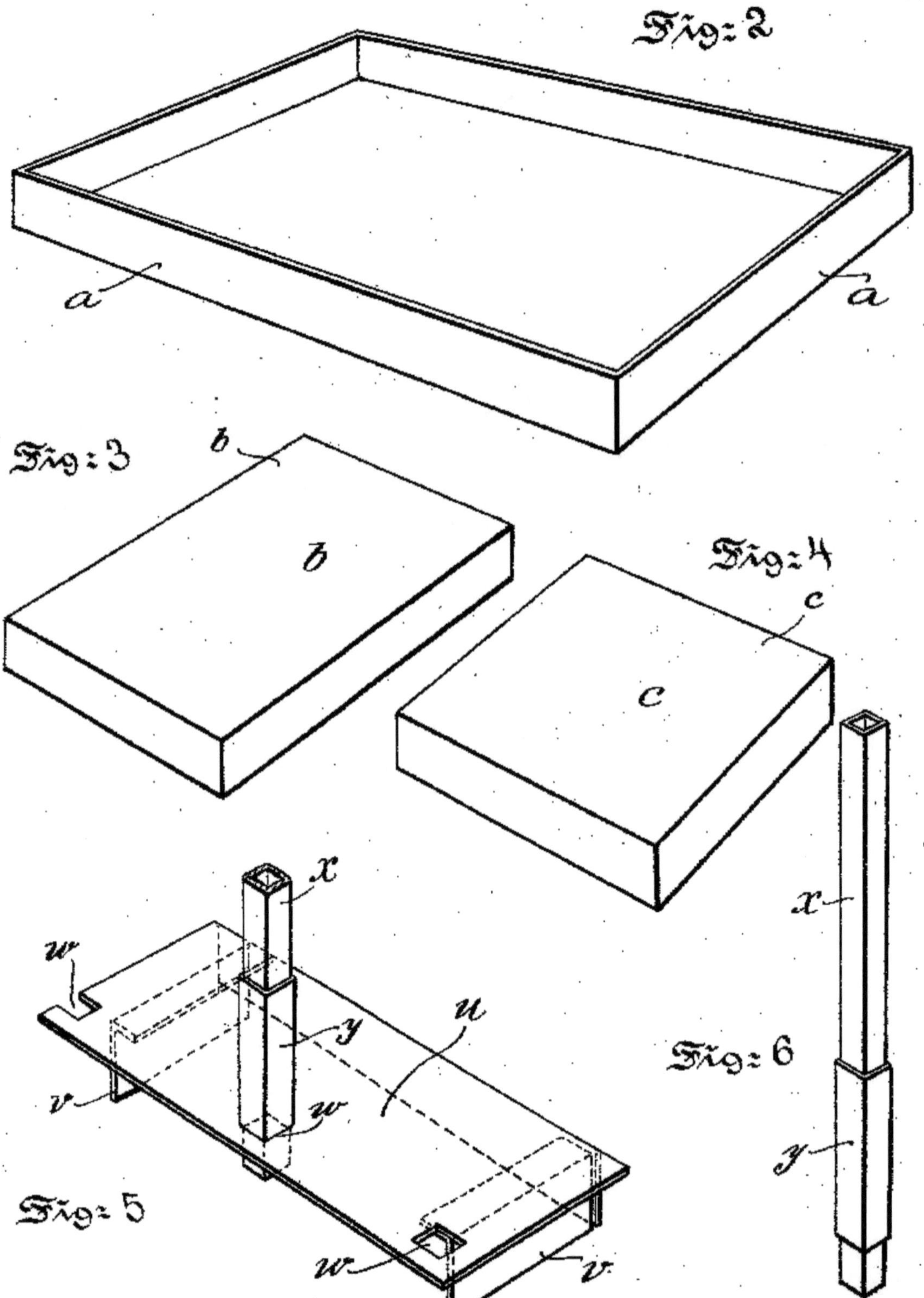

Fig. 2
a
a
Fig. 3
b
b
Fig. 4
c
c
x
w
y
u
v
w
x
Fig. 6
y
Fig. 5
w
v
Witnesses:
Inventor.
Gerhardt E. Grimm
By his Attorney Chas. A. Pinter.

No. 739,669.

PATENTED SEPT. 22, 1903.

G. E. GRIMM.

KNOCKDOWN TOY HOUSE.

APPLICATION FILED JULY 26, 1902

NO MODEL.

6 SHEETS—SHEET 3.

Fig. 7

Fig. 9

Fig. 8

Fig. 10

Witnesses:
W. A. Schafer

Inventor.
Gerhardt E. Grimm.
By his Attorney Chas. A. Rutter.

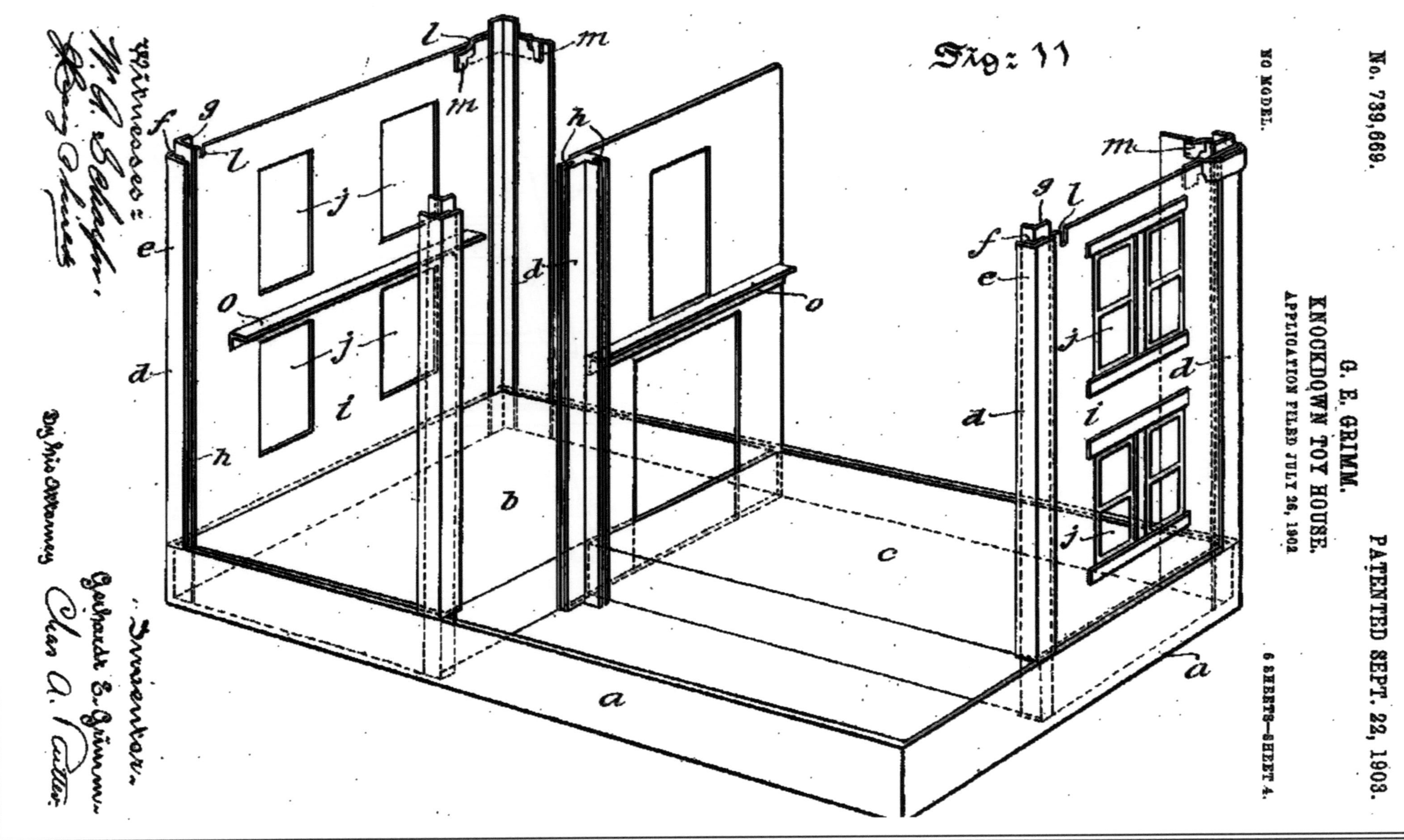
No. 739,669.
PATENTED SEPT. 22, 1903.
G. E. GRIMM.
KNOCKDOWN TOY HOUSE.
APPLICATION FILED JULY 26, 1902.
NO MODEL.
6 SHEETS—SHEET 4.
Fig: 11
Witnesses:
Inventor.
Gerhardt E. Grimm
Chas A. Cutter
By his Attorney

No. 739,669.

PATENTED SEPT. 22, 1903.

G. E. GRIMM.
KNOCKDOWN TOY HOUSE.
APPLICATION FILED JULY 28, 1902.

NO MODEL.

6 SHEETS—SHEET 5.

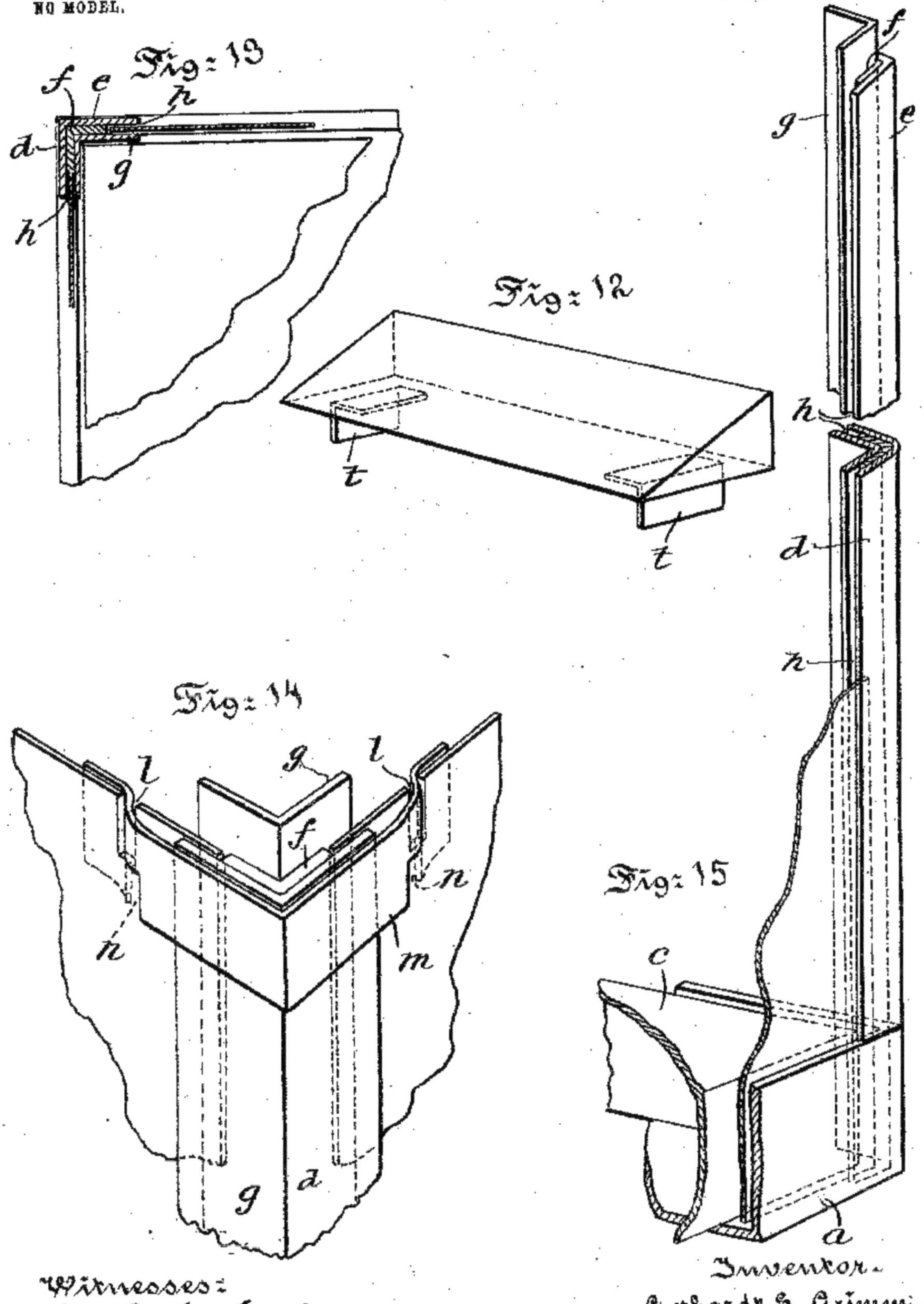

Witnesses:
W. A. Schaffer.

Inventor.
Gerhardt E. Grimm.
By his Attorney Chas. A. Ritter.

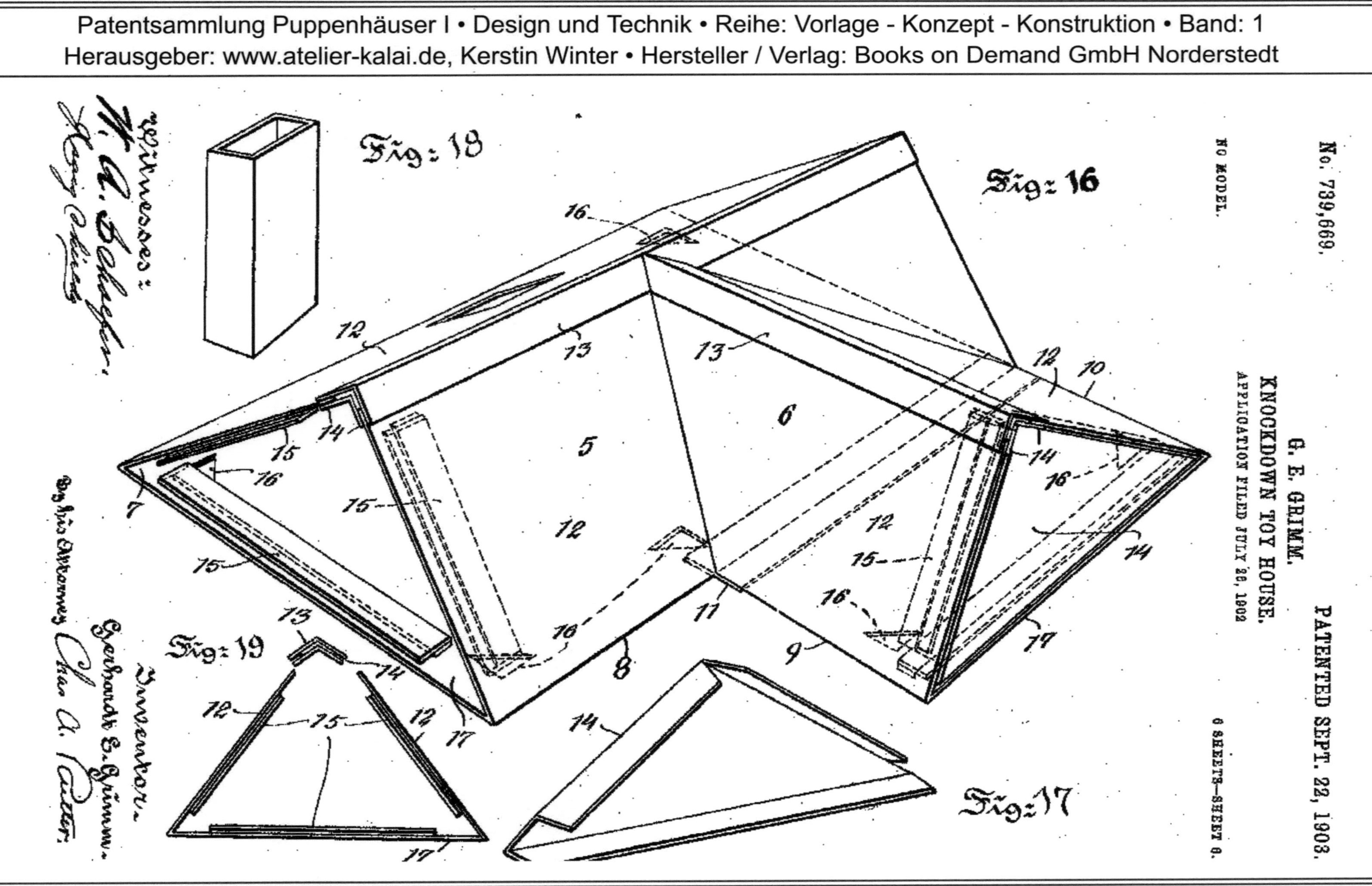
No. 739,669.
PATENTED SEPT. 22, 1903.
G. E. GRIMM.
KNOCKDOWN TOY HOUSE.
APPLICATION FILED JULY 25, 1902.
NO MODEL.
6 SHEETS—SHEET 6.
Fig: 18
Fig: 16
Fig: 19
Fig: 17
Witnesses:
Inventor.
Gerhardt E. Grimm.
By his Attorney Chas. A. Petter.

No. 739,669.

Patented September 22, 1903.

UNITED STATES PATENT OFFICE.

GERHARDT E. GRIMM, OF WOODBURY, NEW JERSEY.

KNOCKDOWN TOY HOUSE.

SPECIFICATION forming part of Letters Patent No. 739,669, dated September 22, 1903.

Application filed July 26, 1902. Serial No. 117,195. (No model.)

To all whom it may concern:

Be it known that I, GERHARDT E. GRIMM, a citizen of the United States, and a resident of Woodbury, Gloucester county, New Jersey, have invented certain new and useful Improvements in Knockdown Toy Houses, of which the following is a specification.

My invention relates to improvements in toy houses; and the object of my invention is to furnish a knockdown toy house of paper which may be erected or taken apart in a few minutes, which will be inexpensive to manufacture, and which will be very durable.

My invention consists, first, in the arrangement of the base, in which a part of the box in which the several pieces from which the house is constructed are packed for shipment or storage, serves to bind or retain the floors of the lower story of the house, and in connection with these floors to form a means for securely supporting the corner-pieces of the house; secondly, in the construction of the corner-pieces of the house, which securely carry the side walls and roof; thirdly, in a novel means for securing the upper corners of the house; fourthly, in the construction of the roof of the house; fifthly, in certain other novel features of construction, which will be hereinafter fully described.

In the accompanying drawings, forming part of this specification, and in which similar characters of reference indicate similar parts throughout the several views, Figure 1 is a perspective view of my toy house when completely erected; Fig. 2, a perspective view of a part of the box in which the pieces of the house are stored; Figs. 3 and 4, perspective views of the pieces for making the floors of the lower story of the house; Fig. 5, a perspective view of the porch-floor and part of one of the posts; Fig. 6, a perspective view of one of the posts for supporting the porch-roof; Fig. 7, a perspective view of one of the side walls of the house, showing window; Fig. 8, a perspective view of bay-window construction; Fig. 9, a perspective view of a side wall of the house, showing windows and doors and means for carrying the porch-roof; Fig. 10, a perspective view showing under and rear side of porch-roof; Fig. 11, a perspective view of the house, the end walls, a partition-wall, and the corner-posts; Fig. 12, a perspective view of the top of the bay-window; Fig. 13, a sectional plan showing corner-post and sides of house held by this post; Fig. 14, a perspective view of upper end of corner-post and upper portions of side walls held by this post, showing means for locking these several pieces together; Fig. 15, a perspective view of corner-post and part of lower floor of house and part of inclosing base, showing how corner-post is carried at its bottom; Fig. 16, a perspective view of the roof, one of the gable ends being removed; Fig. 17, a perspective view of one of the gable ends; Fig. 18, a perspective view of the chimney; Fig. 19, an end elevation of one portion of the roof, showing the ridge-pole removed.

For cheapness and lightness the house is constructed of a strong paper, the so-called "strawboard" being an excellent material. The paper is printed or painted in any manner or color suitable for the purpose. The windows of the house are preferably constructed of mica, isinglass, celluloid, or any other suitable transparent material. They are bound or edged with paper to form a sash or printed for the purpose, and are glued in suitable openings formed in the sides of the house.

As has been before stated, the principal object of my invention is to furnish a toy house which may be taken apart and placed in a box for shipment or storage. A part of the box a, Figs. 2, 11, and 15, in which the pieces of the house are placed for shipment or storage, serves to bind together the pieces which form the floor of the lower story of the house and, in connection with these pieces, to form the means for securing the lower ends of the corners of the house and the lower ends of the sides. In Figs. 3 and 4 the pieces forming the floor of the lower story of the house are shown, b being one piece and c the other, and in Fig. 11 these pieces are shown in place in base a. The pieces b c are made from flat pieces of paper scored along their sides parallel to their edges, so that they may be bent down to form a box-like structure when they are to be used for the flooring or opened out perfectly flat when the house is knocked down for shipment.

d, Figs. 1, 11, 13, 14, and 15, represents the corner-posts of the house. They are constructed of three pieces of paper *e f g*, each piece being bent at right angles along its longitudinal axis and secured one to the other by glue or in some other suitable manner. The central piece *f* is of considerably less width than the inner and outer pieces *g e*, so as to form a groove *h*, which is adapted to receive the edges of the sides of the structure. At the lower and upper ends the piece *g* projects beyond the pieces *e f*. At the lower end this projection is passed in between the inclosing box *a* and the floor-pieces *b c*, which form an anchorage for it, and at the upper end it enters a hole in the pieces forming the roof, as presently described. These projecting ends are dipped in glue or some other suitable material in order to stiffen them and prevent any liability of their fraying out in use.

The sides *i* of the house are constructed from flat sheets of paper, which are punched out to form windows *j*, which are covered with any suitable transparent material to represent glass. The material forming the windows is printed or has paper pasted upon it to represent sash. The edges of the sheets *i* are adapted to be placed within the grooves *h* in the corner-posts, and in order that they may enter these grooves easily they (the sides) should be somewhat thinner than the material out of which the central piece *f* of the post is constructed. The lower ends of the sides are adapted to be passed in between the box *a* and the sides of the flooring *b c* and are thereby securely held in place. Near the corners of the upper ends of the sides *i* I form notches *l*, Figs. 7, 9, 11, and 14, which are adapted to interlock with a fastening *m*, made of paper, to prevent the upper parts of the sides from being withdrawn from the corner-pieces.

As shown in Fig. 14, the fasteners *m* are constructed from a flat piece of paper, having near its ends notches *n*, which are adapted to engage one with the notch *l* upon one side *i* and the other with the notch *l* upon the adjacent side.

For the second story the floors consist of a flat piece of paper which is adapted to be supported by the brackets *o*, carried by the inner sides of the walls *i*. These brackets are preferably made from a flat thick piece of paper scored longitudinally along their centers and glued to the wall. Their unglued or upper portion is adapted to be bent over, as shown in Fig. 11, to support the floor and to be raised up to lie flat against the wall when the several parts are packed in the shipping-box.

The bay-window construction is shown in Fig. 8 and its top in Fig. 12. The bay is made from a flat piece of paper scored so that it may be bent to form sides *p* and wings *r*, bent at right angles to the sides, which are adapted to enter the grooves *h* in the corner-pieces *d*, as shown in Fig. 1. The bottom of the bay is formed by scoring and bending under the lower part of the front, as shown in Fig. 8, and this bottom is further scored and bent to form a tongue *s*, which is adapted to be passed in between the side of the box *a* and the end of the floor *b*. The top of the bay is formed of the closed box-like structure shown in Fig. 12 and has on its lower side fasteners *t*, adapted to engage the inner upper sides *p* of the bay. These fasteners are formed of flat scored pieces of paper, one part adapted to be glued to the top and the other part adapted to be bent down, as shown.

The floor *u* of the porch (shown in Figs. 1 and 5) is formed of a stiff flat piece of paper and is furnished with fasteners *v*, similar in all particulars to the fasteners *t* of the bay-window roof, which hold it in place in the box *a*. It is furnished with perforations *w*, adapted to receive the lower ends of the columns *x*, which are paper tubes of any suitable section. *y* is a collar on tubes *x*, which serves as a stop to hold them on the floor *u* as well as an ornament.

The porch-roof (shown in Figs. 1 and 10) is a box-like structure of triangular section. At its rear end it is furnished with an Γ-shaped fastening 1, which is adapted to engage a similar fastening 2, glued or otherwise secured to the side of the house, as shown in Fig. 9. The piece forming the under side of the porch-roof is furnished with openings 3 to receive the upper ends of the columns *x*.

4, Figs. 1 and 9, is a door opening out on the porch. This door is preferably secured to the wall in which it is placed by a paper hinge of any suitable kind.

The roof of the house is made so that when knocked down it may be folded perfectly flat. It consists, essentially, of two parts—one, 5, covering the main part of the house and the other, 6, the wing. The part 5 is hinged at 7 8 and the part 6 at 9 and 10, and the part 6 is hinged to the part 5 at 11. The hinges may and preferably are formed by simply scoring the paper in a manner well known to paper-workers. When the sides 12 of the roof are raised, they are held by a joining-strip or ridge-pole 13, which has upon its edges grooves 14 to receive the ends of the sides. The joining-strip is formed of three pieces of paper glued together in a manner similar to those forming the corner-posts *d*. The gable ends of the roof are formed of triangular box-like structures 14, which are passed into the ends of the roof, and the sides of which are engaged by Γ-shaped locking-pieces 15, glued to the roof structure. These locking-pieces not only hold the end pieces in place, but in connection with these pieces insure the roof keeping its shape when raised.

16 represents perforations in the bottom part 17 of the roof, through which the tops of the corner-posts *d* pass to hold the roof on the sides.

Having thus described my invention, I claim as new and desire to secure by Letters Patent—

1. In a toy house, in combination, a base having a bottom and sides, a floor for each lower room consisting of a box-like structure having a top and sides, corner-posts the lower ends of which are adapted to be held between the sides of the base and the sides of the flooring, and the edges of which are furnished with vertical grooves, sides adapted at their ends to be held by the grooves in said posts and at their bottoms between the sides of the base and the floors, and a roof adapted to be removably held by the upper ends of said corner-posts.

2. In a toy house in combination, a base having a bottom and sides, a floor for each lower room consisting of a separate box-like structure, adapted to be held in said base, corner-posts the lower ends of which are adapted to be held between the sides of the base and the sides of the flooring, and the edges of which are furnished with vertical grooves, sides adapted at their ends to be held by the grooves in said posts, and at their bottoms between the sides of the box and the floors, partition-walls separating the rooms adapted at their bottoms to be held between the box-like structures forming the lower floors, and a roof adapted to be removably carried by the upper ends of said corner-posts.

3. The described means for holding the lower ends of the side walls and partitions of a knockdown toy house, comprising in combination with said walls and partitions, a base having a bottom and sides, and separate box-like structures held in place by said base and forming each a floor for a lower room of said house.

4. In a toy house, in combination, a base having a bottom and sides, separate box-like structures held in place by said base and forming each a floor for a lower room of said house, side walls adapted at their lower ends to be held between the sides of said base and the sides of said floors, a partition-wall adapted to be held at its lower end between the sides of said floors, means for securing together adjoining corners of said side walls, means carried by said side walls and partitions for carrying upper floors, said upper floors, and a detachable roof.

5. In a collapsible toy house, a base having bottom and sides, floor-sections removably mounted in said base, and walls removably fitted between the sides of the base and floor-sections respectively.

6. In a collapsible toy house, a base having a bottom and sides, floor-sections removably mounted in said base and each consisting of a box-like structure having a top and sides bent at right angles thereto, and walls removably fitted between the sides of the base and floor-sections respectively.

7. In a collapsible toy house, a base composed of a box-like structure constituting the packing-box for the house, floor-sections removably mounted in said base, walls removably fitted between the sides of the base and floor-sections respectively, and a collapsible roof removably supported upon the walls.

8. In a collapsible toy house, a base having a bottom and sides, floor-sections removably mounted in said base, walls removably fitted between the sides of the base and floor-sections respectively, corner-posts projecting above the house-walls and a roof provided with openings to receive the projecting ends of the corner-posts.

9. In a toy house, in combination, a house structure as described, a porch-floor formed from a flat sheet of paper adapted to be removably carried by the base and furnished with perforations to receive columns, said columns, a box-like roof of triangular cross-section furnished on its under side with perforations to receive the tops of said columns, an Γ-shaped fastening carried by the rear side of said roof, and an Γ-shaped fastening carried by the wall of said house and adapted to engage the fastening carried by the roof.

10. In a collapsible toy house, in combination, a base having a bottom and sides, floor-sections removably supported within the base, corner-posts removably supported between the sides of the base and the floor-sections at the corners thereof, house-walls supported by said base and floors and corner-posts, and a roof provided with openings to receive the upper ends of the corner-posts.

11. The combination with a toy house of a collapsible bay-window having a front and sides constructed from a flat piece of paper scored and bent as described so as to be collapsed, means for securing said window to one of the side walls of the house, and a top or roof covering the top of said bay-window.

12. In a collapsible toy house, a collapsible bay-window having side wings, and means for attaching said wings to the house.

13. In a collapsible toy house, a collapsible bay-window attached thereto and a roof removably supported upon the bay-window.

14. In a collapsible toy house, a wall, a collapsible bay-window attached thereto, a roof removably supported upon said bay-window, and means for holding the roof in position.

15. In a collapsible toy house, a collapsible bay-window formed from a strip of scored material having means for attachment to the house and a roof having downwardly-projecting wings adapted to enter the open top of the bay-window whereby to support the roof in position, and prevent the window from collapsing.

16. In a collapsible toy house, a base, walls removably supported upon the base, grooved corner-posts removably fitted over the corners of the walls and a bay-window having side wings constructed to enter grooves in the posts and support the bay-window in position.

17. The combination in a roof for toy houses, of a flat piece of heavy paper having hinged sides, a grooved joining-strip adapted to secure the upper ends of said sides, and box-

like gable-pieces for closing the ends of the roof.

18. In a collapsible toy house, a house-wall, a fastening-piece 2 having one edge secured thereto and its other edge free and a porch-roof having a similar fastening-piece secured to the rear side thereof and arranged to interlock with the fastening-piece on the house-wall.

19. In a collapsible toy house, a base, a porch-floor removably supported thereon, a house-wall having a fastening-strip 2 secured to the face of said wall, a porch-roof having a similar fastening-piece secured to the rear side thereof and arranged to interlock with the said fastening-piece on the house-wall and posts having their opposite ends removably inserted into openings in the said floor and roof respectively.

20. The combination in a roof for toy houses, of a flat piece of heavy paper having hinged sides, T-shaped fasteners carried by the inner parts of the said sides, a grooved joining-strip adapted to secure the upper ends of said sides, and a box-like gable-piece for closing the ends formed by the sides and bottom pieces of the roof, the sides of said gable-piece being adapted to pass under the fasteners carried by the roof.

21. In a collapsible toy house, a collapsible gable-roof comprising a body portion which constitutes the ceiling of a room and side portions hinged at their outer edges to the body portion and adapted to fold thereon and a ridge-piece detachably fitted to the adjacent edges of said side portions.

22. In a collapsible toy house, a gable-roof removably fitted thereto and comprising a body portion, side portions hinged at their outer edges to said body portion and adapted to fold thereon and a grooved ridge-piece into which the adjacent edges of the side portions fit.

23. In a collapsible toy house, a collapsible gable-roof, flanged end pieces removably fitted in the open ends of the gable, and locking-pieces secured to the said roof and coöperating with the flanges of said end pieces to hold the latter in position.

24. In a collapsible toy house, a collapsible roof, consisting of a main section comprising a body portion and side pieces hinged thereto and a second roof-section arranged at right angles to the main roof-section and having a hinged connection therewith, to permit one section to fold over upon the other.

25. In a collapsible toy house, a base, comprising a box-like structure for containing the various parts of the house when collapsed, floor-sections removably fitted in said base, house-walls removably fitted between the sides of the floor-sections and base respectively, and a partition having its lower end removably fitted between the adjacent ends of the said floor-sections.

26. In a collapsible toy house, a base, a roof, walls removably supported upon the base, and floor-sections removably supported by said house-walls between the roof and the base, whereby the house is divided into upper and lower stories.

27. In a collapsible toy house, a base, house-walls removably supported upon the base, supporting pieces or brackets attached to the interior of the house-walls and floor-sections removably supported upon said supporting pieces or brackets.

28. In a collapsible toy house, a base, walls removably supported upon the base, a vertical partition dividing the house into separate rooms, brackets secured to the house-walls and to opposite sides of the partition and floor-sections removably supported upon said brackets.

GERHARDT E. GRIMM.

Witnesses:
A. STANLEY PETERSON,
CHARLES A. RUTTER.

July 12, 1932.   H. L. MYERS   1,867,374

TOY HOUSE

Filed May 1, 1931   4 Sheets—Sheet 1

Fig.1.

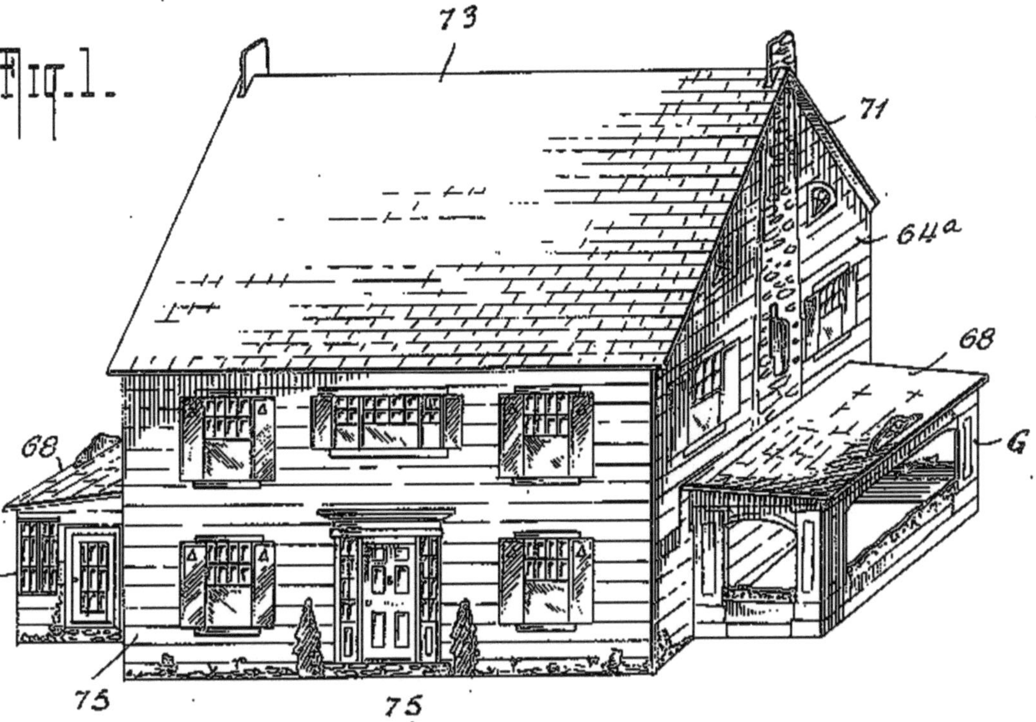

Fig.2.

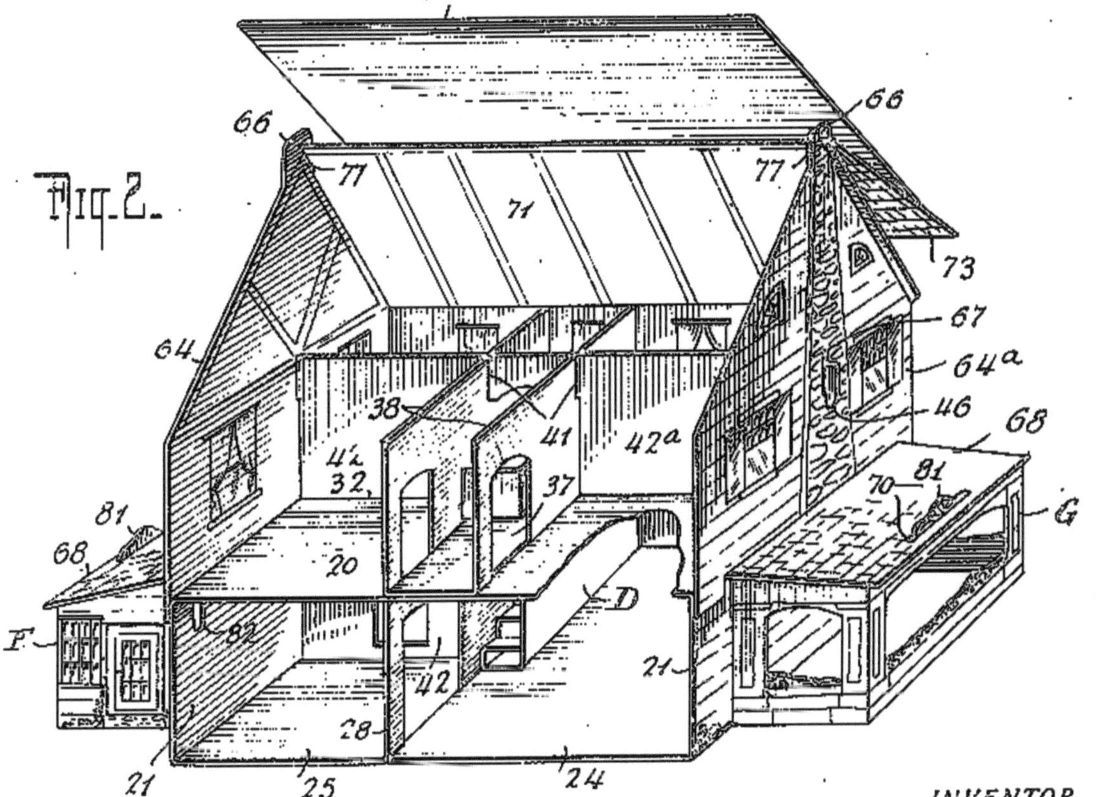

INVENTOR
HAROLD L. MYERS
BY
ATTORNEYS

H. L. MYERS

TOY HOUSE

1,867,374

Filed May 1, 1931

4 Sheets—Sheet 2

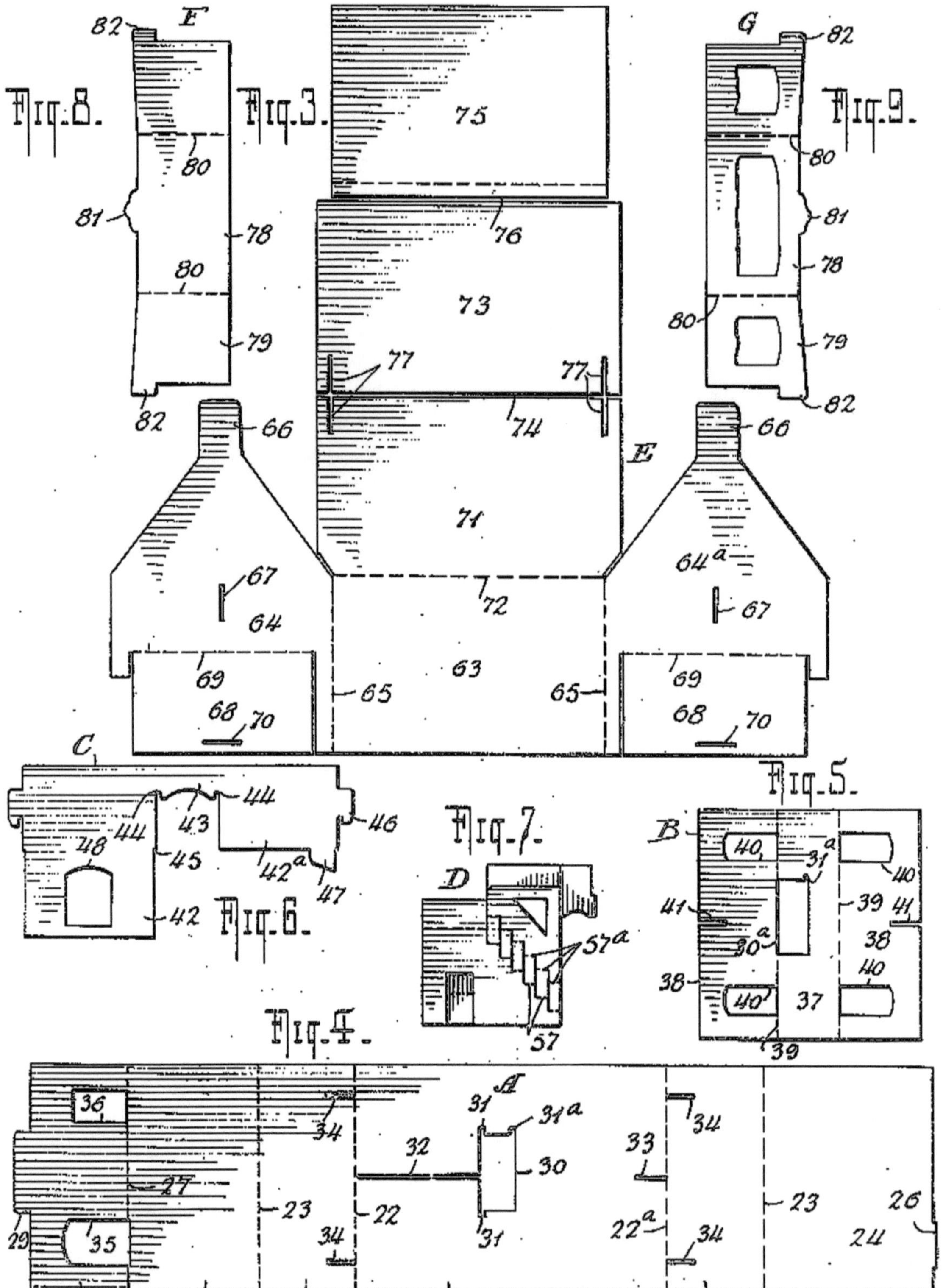

Fig.8. Fig.3. Fig.9.

Fig.5.

Fig.7.

Fig.6.

Fig.4.

INVENTOR
HAROLD L. MYERS

BY

ATTORNEYS

Patentsammlung Puppenhäuser I • Design und Technik • Reihe: Vorlage - Konzept - Konstruktion • Band: 1
Herausgeber: www.atelier-kalai.de, Kerstin Winter • Hersteller / Verlag: Books on Demand GmbH Norderstedt

July 12, 1932.

H. L. MYERS

1,867,374

TOY HOUSE

Filed May 1, 1931

4 Sheets—Sheet 3

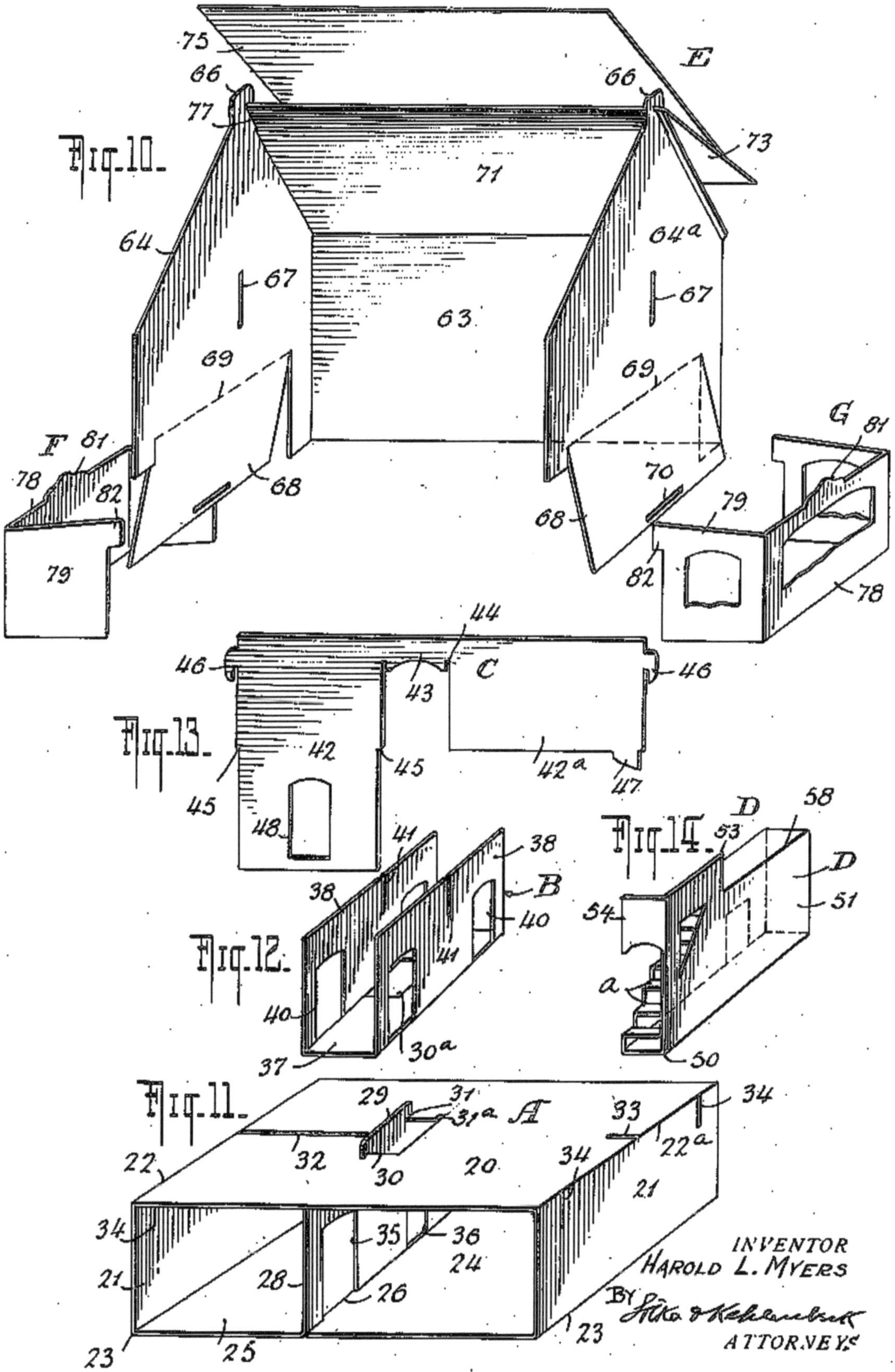

Fig.10.

Fig.13.

Fig.12.

Fig.14.

Fig.11.

INVENTOR
HAROLD L. MYERS
BY
ATTORNEYS

July 12, 1932.　　　H. L. MYERS　　　1,867,374

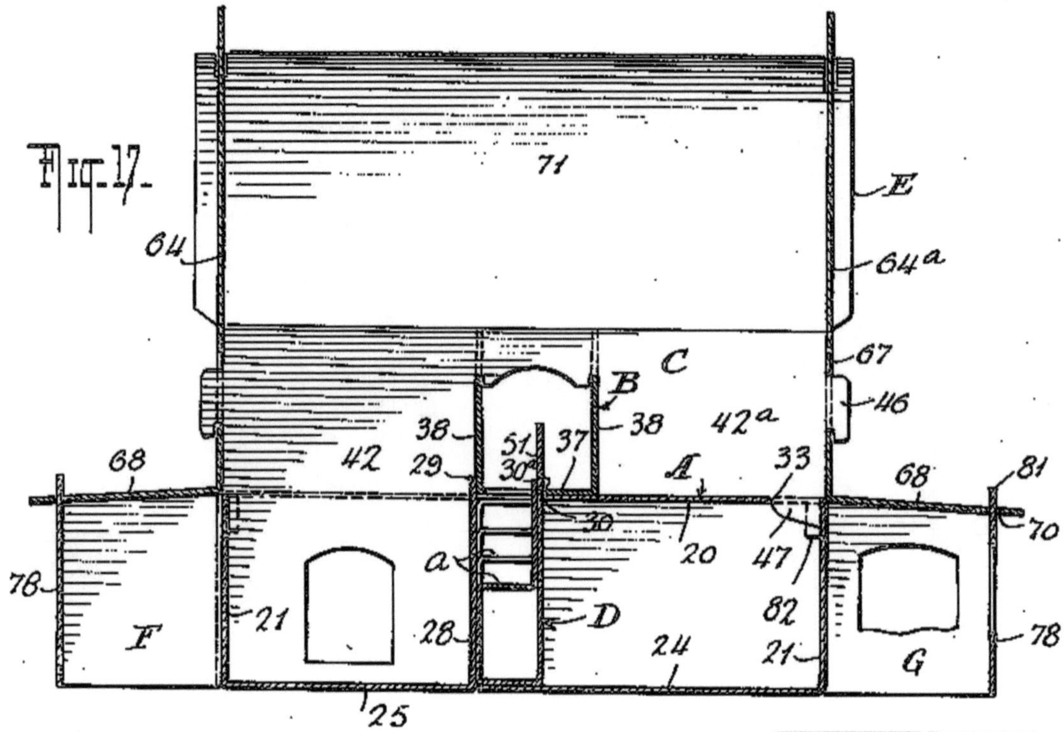

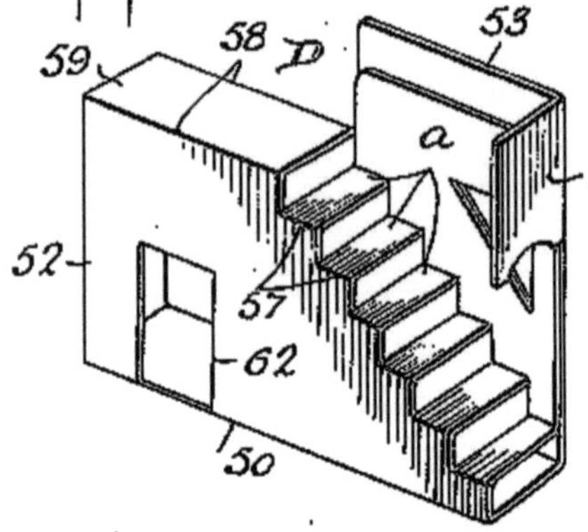

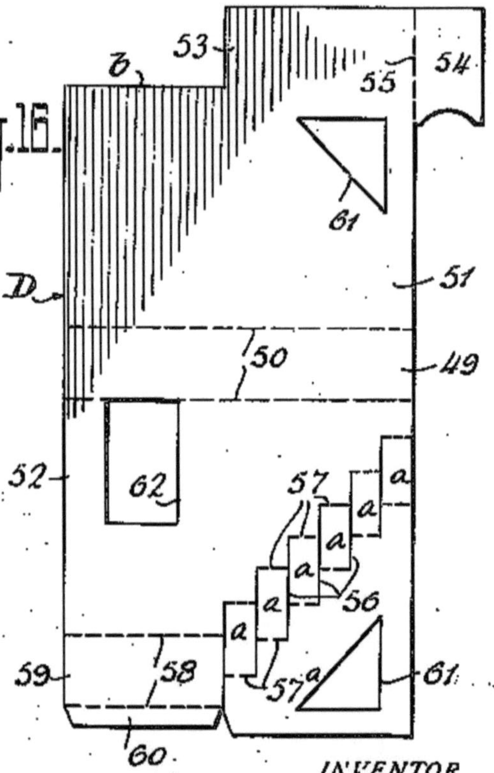

TOY HOUSE

Filed May 1, 1931　　　4 Sheets-Sheet 4

Patented July 12, 1932

1,867,374

UNITED STATES PATENT OFFICE

HAROLD L. MYERS, OF MORRISTOWN, NEW JERSEY

TOY HOUSE

Application filed May 1, 1931. Serial No. 534,197.

My invention relates to toy houses and more particularly to that type thereof commonly known as doll houses, and has for its object to construct such a house of cardboard or equivalent material, in a manner to permit the house to be folded or collapsed into compact form when not in use, and to be easily set up and assembled to constitute a replica of a conventional cottage or other building. The invention contemplates further the provision of a toy house in which a plurality of rooms or equivalent divisions are contained all of which are accessible for play purposes. Other objects will appear from the description hereinafter, and the features of novelty will be pointed out in the claims.

In the accompanying drawings, which illustrate an example of the invention without defining its limits, Fig. 1 is an exterior perspective elevation of a doll house embodying the novel features; Fig. 2 is a perspective view showing the roof and one wall of the house in a folded position in which the interior parts of the house are accessible; Figs. 3 to 9 inclusive are views of the blanks or sections which form the several parts of the house; Figs. 10 to 14 inclusive are views showing said blanks or sections in positions ready to be connected to set up the house; Fig. 15 is a perspective view of a folding stairway forming part of the invention; Fig. 16 is a face view of the blank from which said stairway is constructed; and Fig. 17 is a longitudinal section of the house shown in Fig. 1.

In the illustrated example the toy house is shown in the form of a conventional dwelling, and consists of a plurality of sections of cardboard or other suitable material capable of being folded into shape and fitted together to form the several floors, rooms and other divisions included in the dwelling. While the novel arrangement is particularly adapted for the construction of toy dwellings, it is not restricted thereto and the drawings and following description are accordingly to be construed as including equivalent combinations which, when set up, constitute replicas or other types and classes of buildings.

The main or lower portion of the house is formed from the blank or section A shown in Figs. 4 and 11, which consists of a main panel 20 and auxiliary panels 21 foldably connected therewith along score lines 22 and 22ᵃ; the panels 21 in turn are foldably connected along score lines 23 with additional panels 24 and 25, the panel 24 being provided at its free edge with a projection 26 and the panel 25 being foldably connected along a score line 27 with a partition panel 28 having a tab 29 projecting from its free edge, as shown in Fig. 4. The main panel 20 is formed with an opening 30 which, in the finished house becomes a stairwell, and is provided with notches 31 and 31ᵃ, the purpose of which will be more fully referred to hereinafter. In addition to the above, a slot 32 is cut in the main panel 20 so as to extend from the opening 30 to the score line 22, said slot serving to accommodate a partition member in the manner to be set forth in detail further on in the description. A short slot 33 is cut in the main panel 20 in registry with the slot 32 and extending from the score line 22ᵃ, while additional short slots 34 are cut in the panels 21 so as to project from the score lines 22 and 22ᵃ in registering pairs as shown in Fig. 4. It will be understood that the arrangement of the slots 32, 33 and 34 may be changed to meet the requirements of the building which is to be exemplified by the toy house and in some cases may be all or partly omitted. In the present case the partition panel 28 is provided with a main opening 35 and a minor opening 36 both representing door openings in the finished house, the opening 35, in the illustrated example, being in linear registry with the projection 26 of the panel 24.

When setting up the toy house, the section A is folded to the condition shown in Fig. 11; that is, the auxiliary panels 21 are folded on the score lines 22 and 22ᵃ to positions at right angles or perpendicular to the main panel 20 and the panels 24 and 25 are folded on the score lines 23 into parallel relation with the main panel 20, the partition panel 28 having been folded on the line 27 to a position perpendicular to the panel 25. The tab 29 is passed through the opening 30 and fitted into the notches 31 and the projection 26 is fitted into the lower end of the door opening 35 so that the parts are firmly held in the form of a rectangular tube divided transversely by the partition panel 28.

In the present case, the upper floor of the house is divided into a hall and several rooms by means of a cardboard hall section B shown in Figs. 5 and 12. This section B consists of a central panel 37 and two side panels 38 foldably connected therewith along score lines 39. The central panel 37 is provided with an opening 30ᵃ having a notch 31ᵃ corresponding to and adapted to register with the opening 30 and the notch 31ᵃ of the section A; the side panels 38 in turn include door openings 40 and slits 41, the latter terminating at the free edges of said panels 38 and being located so as to register lineally with the slots 32 and 33 of the section A.

In setting up the toy house, the section B is folded to the position shown in Fig. 12, in which the side panels 38 extend at right angles to the panel 37, and in this position is set upon the section A with the opening 30ᵃ in registry with the opening 30.

The upper floor of the house chosen for purposes of illustration is further divided by a partition section C shown in Figs. 6 and 13. This section C consists of a flat sheet of cardboard or other suitable material cut to form a major member 42 and a minor member 42ᵃ connected by a cross member 43, preferably of ornamental outline and provided with notches 44. At intermediate points in registry with each other, the major member 42 is formed with shoulders 45 and further includes a locking hook or tab 46 which projects from the free edge of said member 42 and is duplicated at the free edge of the member 42ᵃ; a tab 47 projects from the lower edge of the member 42ᵃ, and a door opening 48 is provided in the member 42, as shown in Figs. 6 and 13.

When the toy house is to be set up, the section C is put in place by passing the member 42 downwardly through the slit 30 until the shoulders 45 arrest such movement; in this position the member 42 extends between the partition panel 28 and the one end panel 21 and divides the lower floor of the toy house into two rooms on one side of said partition panel 28, which rooms communicate with each other through the door opening 48, as shown in Fig. 2; at the same time the member 42ᵃ and the upper portion of the member 42 divide the upper floor of the house into additional rooms on each side of the hall section B. When the section C is fully in place, the cross member 43 will be located in the notches 41 of the section B and the latter will be fitted into the notches 44 of the section C; at the same time, the tab 47 will project into the slot 33 of the section A so that the sections A, B and C are firmly fixed in place.

The toy house in its illustrated form further includes a collapsible stair section D shown in Figs. 7, 14, 15 and 16. The blank form in which this stairway is constructed is shown in Fig. 16 and consists of a sheet of cardboard or other suitable material cut to form a bottom panel 49 foldably connected along score lines 50 with staircase panels 51 and 52. The panel 51 is continued to provide a rail member 53, which includes a rail flap 54 foldable on a score line 55 for the purpose to be more fully set forth hereinafter, while the panel 52 is formed with a plurality of spaced slits 56 located in staggered relation and having their opposite ends terminating at score lines 57 and 57ᵃ, which extend between each pair of slits 56 to provide a plurality of stair panels a, as shown in Fig. 16. Spaced parallel score lines 58 extend from the one terminal slit 56 to the free edge of the panel 52 to form an auxiliary panel 59 and a glue flap 60 foldable on said score lines 58. To increase the effect of a bannister the panels 51 and 52 may be provided with corresponding openings 61 of triangular or other predetermined shape adapted to register with each other in the assembled condition of the stair section D. In the illustrated example the panel 52 is further provided with a door opening 62.

To bring the stair section D into condition for use in the toy house, the panel 52 is folded on its score line 50 into surface engagement with the panel 51 with the openings 61 in registry with each other, as shown in Fig. 7. That part of the panel 52 which includes the opening 61 is then glued to the panel 51, the gluing extending up to the score lines 57ᵃ, but not beyond the same to the stair panels a; at the same time, the glue flap 60 is folded over the edge b of the panel 51 and glued to the rear surface thereof.

In the setting up of the toy house, the stair section D is placed in position by first folding back the rail flap 54 and then passing the rail member 53 and folded flap 54 from below upwardly through the stairwell formed by the openings 30 and 30ᵃ, the rail member 53 fitting into the notches 31ᵃ. When this upward insertion has been completed the section D is unfolded on the score lines 50, 57 and 57ᵃ, and 58, to form the flight of stairs shown in Figs. 14 and 15, and the rail flap 54 is folded to the position shown in said figures and so as to extend between the rail member 53 and one side panel 38 of the section B. In the final position of the section D the stairway rises adjacent to and in contact with the partition panel 28 of the section A, and the panel 49 lies beneath and in engagement with the panel 20 of said section A. The effect is thus one of a rising flight of stairs and a rearwardly extending hall which communicates with one of the rooms on the lower floor of the house through the door opening 61.

In addition to the parts so far described, the toy house includes a section E illustrated in Figs. 3 and 10 and consisting of a rear wall panel 63 and side wall panels 64 and 64ª foldably connected with the panel 63 along score lines 65; the side wall panels 64 and 64ª which constitute the side walls of the toy house taper upwardly to define the roof lines and terminate in upwardly extending projections 66 shaped to represent chimneys. The side wall panels 64 are provided with vertical slots 67, and are further slit to form porch roof panels 68 foldable on score lines 69, said panels 68 including horizontal slots 70, as shown in Figs. 3 and 10. The roof of the toy house consists of a panel 71 foldably connected with the panel 63 along a score line 72, and a panel 73 foldably connected with the panel 71 along a folding line 74; the latter is preferably taped to provide a hinge on which the panel 73 may be folded back upon the panel 71 for the purpose to be more fully set forth hereinafter. A front wall panel 75 is foldably connected with the panel 73 along a folding line 76, which preferably is also taped to provide a hinge on which the panel 75 may be folded upon the panel 73. The panels 71 and 73 are slotted at 77 to provide slots extending in opposite directions beyond the hinge line 74 for the reception of the chimney projections 66 in the completely set up condition of the house.

In setting up the house, the section E is folded about the section A to bring the rear wall panel 63 across the back thereof and the side wall panels 64 and 64ª into positions at right angles to said panel 63 in engagement with the end panels 21 of said section A, the locking hooks 46 of the section C being passed through the slots 67 to lock the section E in place. The roof panels 71 and 73 are then folded over and into engagement with the upper edges of the side wall panels 64 and 64ª; the chimney projections 66 extending through the slots 77, and the front wall panel 75 automatically folding into place to close the front of the house and complete the main structure thereof.

The illustrated form of the toy house further includes two outside end sections F and G shown in Figs. 8, 9 and 10 and constructed to represent a sun porch and an open porch respectively. The sections F and G comprise central panels 78 and end panels 79 foldably connected with the central panels 78 along score lines 80, the central panels being each provided with lugs 81 and the side panels with projections 82.

The sections F and G are placed in position beneath the porch roof panels 68 by folding the end panels 79 to the positions shown in Fig. 10 and inserting the projections 82 into the slots 34 of the section A; the roof panels 68 are then folded down upon the sections F and G with the lugs 81 projecting through the slots 70 to complete the structure.

It will be understood that all of the various sections constituting the house are suitably decorated exteriorly and interiorly to carry out the intended effect, for instance, as shown in Figs. 1 and 2, in which the dwelling includes two end porches, and interiorly is divided into a living room, dining room and kitchen on the first floor and three bed rooms, a bathroom and a centre hall on the second floor, with a stair case giving access from the living room on the first floor to said hall on the second floor.

To make the setting up of the toy house clear and easily understood, the following brief resumé will be of assistance. The first step is to fold and secure the section A to the position shown in Fig. 11, after which the sections B and C in the manner indicated in Figs. 12 and 13 are connected with the section A; at this stage it is preferred to set the stair section D in place by passing it upwardly through the stair well 30, 30ª in the manner previously set forth. The section E is then folded about the combined sections A, B, and C and fixed in place by means of the locking hooks 46 and slots 67, after which the roof panels 71 and 73 and the front wall panel 75 of the section E are folded into position. The setting up of the house is completed by adding the porch sections F and G and fixing them in their intended position by means of the panels 68 with the co-operation of the slots 70 and the lugs 81, and the projections 82 with the slots 34. In this condition, the house of the illustrated example presents the appearance shown in Fig. 1.

When access to the interior of the house is desired, the panels 75 and 73 are folded back to the position shown in Figs. 2 and 10, which may be readily done and does not interfere with the rigidity of the house structure. In this position of the parts, the house presents the appearance shown in Fig. 2 in which it is clearly indicated that all of the rooms included in the dwelling on both floors are readily accessible for play purposes and for the introduction of suitable and appropriate toy furniture. By simply folding the panels 75 and 78 back to their normal positions, the toy house is quickly restored to the closed condition shown in Fig. 1.

From the above, it is clear that the house is firmly locked together in its assembled condition and is safe against any of the ordinary impacts developed during play; at the same time, the house is capable of being quickly dismantled and knocked down to compact form for shipping and storage purposes, the parts of said house, when dismantled, all being flat units which are easily handled and stored. The house in whatever form it may be constructed is inexpensive and attractive to the child as a plaything, and comprises in-

4   1,867,374

terlocking and easily assembled sections which co-operate to provide a solid, rigid structure.

It will be noted that the top panel of the section A constitutes the ceilings of the rooms on the first floor, and the floors of the rooms on the second story of the house, thus reducing the arrangement to the greatest simplicity. The use of portions of the sides of the house to provide roofs for the two end porches, and the collapsible stairway and bannister construction are novel features which add to the simplicity of the construction. This is true also of the folding roof and front wall arrangement whereby access to all of the rooms and all interior parts of the house is greatly facilitated.

Various changes in the specific forms shown and described may be made within the scope of the claims without departing from the spirit of the invention.

I claim:—

1. A collapsible toy house comprising a plurality of collapsible interior sections detachably connected with each other to form the interior divisions of the house, an external collapsible section, including foldable side walls foldable about said interior sections to form the exterior of the house, a collapsible porch section detachably connected with said interior sections to simulate a porch, and a porch roof panel forming part of one of said foldable side walls and foldable out of the plane thereof and relatively thereto into connection with said porch section to form the roof thereof.

2. In a collapsible toy house, an interior section consisting of a plurality of panels foldably connected with each other and adjustable relatively to each other to form a partitioned rectangular horizontal tube adapted to provide the interior floors, ceiling and room walls, and at least a portion of the outer walls of the toy house, and means comprising inherent parts of said panels whereby the latter are detachably locked in position.

3. In a collapsible toy house, an interior section consisting of a main panel arranged to divide the house into an upper and lower floor and provided with a notched opening simulating a stairwell, auxiliary panels foldably connected with said main panel and foldable to positions perpendicular thereto to constitute parts of the side walls of the house, additional panels foldably connected with said auxiliary panels and foldable into parallelism with said main panel to form the bottom floor of the house, a partition panel foldably connected with one of said additional panels and foldable to a position at right angles thereto, said partition panel extending between said additional and main panels, to partition the lower floor of the house into rooms, and a tab on said partition

panel arranged to project through and fit said notched opening for fixing said panels in position.

4. In a collapsible toy house, an interior section consisting of a main panel constituting a ceiling and provided with a notched opening simulating a stairwell, auxiliary panels in permanent hinged connection with said main panel and foldable to positions perpendicular thereto to constitute upright walls, additional panels foldably connected with said auxiliary panels and foldable into parallelism with said main panel to form the floors of said interior section, a partition panel foldably connected with one of said additional panels and foldable to a position at right angles thereto to constitute a room partition, and a tab on said partition panel arranged to project through and fit said notched stairwell opening for fixing said panels in position, said main panel including a slot extending from said notched opening to the folding line of one of said auxiliary panels, a hall section resting upon said main panel and consisting of a central panel and two side panels foldably connected therewith and foldable to positions at right angles to said central panel, the latter being provided with a stairwell opening adapted to register with the stairwell opening of said main panel, and said side panels having slits terminating in open ends at the free edges thereof, and a partition section adapted to be fitted into the slits of said hall section and to extend on opposite sides thereof to provide room divisions above said interior section, said partition section being arranged to project downwardly through the slot of said main panel to form a transverse partition beneath the latter between said partition panel and one of said auxiliary panels whereby said interior section both above and below said main panel is subdivided to provide upper and lower rooms, and a collapsible stair section arranged to be fitted into said stairwell to simulate a stairway communicating with the space above said interior section.

5. In a collapsible toy house, an interior section consisting of a main panel constituting a ceiling and provided with a notched opening simulating a stairwell, auxiliary panels in permanent hinged connection with said main panel and foldable to positions perpendicular thereto to constitute upright walls, additional panels foldably connected with said auxiliary panels and foldable into parallelism with said main panel to form the floors of said interior section, a partition panel foldably connected with one of said additional panels and foldable to a position at right angles thereto to constitute a room partition, and a tab on said partition panel arranged to project through and fit said notched stairwell opening for fixing said panels in position, said main panel including

Patentsammlung Puppenhäuser I • Design und Technik • Reihe: Vorlage - Konzept - Konstruktion • Band: 1
Herausgeber: www.atelier-kalai.de, Kerstin Winter • Hersteller / Verlag: Books on Demand GmbH Norderstedt

a slot extending from said notched opening to the folding line of one of said auxiliary panels, a hall section resting upon said main panel and consisting of a central panel and two side panels foldably connected therewith and foldable to positions at right angles to said central panel, the latter being provided with a stairwell opening adapted to register with the stairwell opening of said main panel, and said side panels having slits terminating in open ends at the free edges thereof, and a partition section adapted to be fitted into the slits of said hall section and to extend on opposite sides thereof to provide room divisions above said interior section, said partition section being arranged to project downwardly through the slot of said main panel to form a transverse partition beneath the latter between said partition panel and one of said auxiliary panels whereby said interior section both above and below said main section is subdivided to provide upper and lower rooms, and a collapsible stair section arranged to be fitted into said stairwell to simulate a stairway communicating with the space above said interior section, locking tabs projecting from said partition section, an exterior section consisting of a rear wall panel and side wall panels foldably connected therewith and provided with slots adapted to fit over the locking tabs of said partition section to enclose said interior section and form the exterior rear and side walls of said house, chimney projections on said side wall panels, roof panels foldably connected with said rear wall panel and with each other and provided with slots arranged to fit over said chimney projections, and a front wall panel foldably connected with said roof panels and arranged to form the front wall of the house.

6. In a collapsible toy house, a collapsible exterior section consisting of a plurality of foldably connected panels arranged to form the side walls, front and rear walls, and roof of said house, an independent porch section consisting of a plurality of foldably connected panels adapted for attachment to a predetermined part of the house to simulate a porch, and a porch roof panel constituting a part of one of the upright walls of the house and foldable relatively thereto for connection with said porch section to form the roof thereof.

7. In a collapsible toy house, a stair blank consisting of a bottom panel, stair case panels foldably connected therewith, and continued to form bannister panels, a rail member forming part of one of said staircase panels, a rail flap foldably connected with said rail member, the other staircase panel being provided with a plurality of parallel slits in staggered relation and folding lines extending transversely between said slits to constitute a flight of stairs, an auxiliary panel foldably connected with said other staircase panel, and

a glue flap foldably connected with said auxiliary panel.

8. In a collapsible toy house, a stair section comprising stair case panels connected with each other in surface engagement along a line defined by score lines arranged in staggered relation opposite to and spaced from correspondingly arranged score lines, one of said panels being formed with a plurality of parallel slits extending between said score lines to define the treads of a flight of stairs, a bottom panel, and an auxiliary panel both connected with said stair case panels along score lines, said panels being foldable relatively to each other on said score lines to adjust said treads into the form of a stair case.

9. In a collapsible toy house, a stair section comprising stair case panels connected with each other in surface engagement along a line defined by score lines arranged in staggered relation opposite to and spaced from correspondingly arranged score lines, one of said panels being formed with a plurality of parallel slits extending between said score lines to define the treads of a flight of stairs, a bottom panel, and an auxiliary panel both connected with said stair case panels along score lines, said panels being foldable relatively to each other on said score lines to adjust said treads into the form of a stair case, the connected portions of the panels constituting bannister and rail members, and a rail flap foldably connected with said rail member.

In testimony whereof I have hereunto set my hand.

HAROLD L. MYERS.

United States Patent [19]

Walmer et al.

[11] **4,216,608**

[45] **Aug. 12, 1980**

[54] **DOLL HOUSE**

[76] Inventors: Harry E. Walmer, 721 N. Overlook Dr., Alexandria, Va. 22305; Judd Horbaly, 125 Commonwealth Ave., Alexandria, Va. 22309

[21] Appl. No.: 3,519

[22] Filed: Jan. 15, 1979

[51] Int. Cl.² A63H 33/52
[52] U.S. Cl. 46/19
[58] Field of Search 46/13, 19, 20, 21, 12; 272/11

[56] **References Cited**

U.S. PATENT DOCUMENTS

739,669	9/1903	Grimm	46/21
1,569,066	1/1926	Beiger	46/19
3,906,659	9/1975	Walmer	46/19
3,996,693	12/1976	Walmer	46/19
4,018,001	4/1977	Walmer	46/19
4,021,960	5/1977	Walmer	46/19
4,094,090	6/1978	Walmer	46/19

FOREIGN PATENT DOCUMENTS

640199 7/1950 United Kingdom 46/21

Primary Examiner—Louis G. Mancene
Assistant Examiner—Wenceslao J. Contreras
Attorney, Agent, or Firm—James Creighton Wray

[57] **ABSTRACT**

The invention relates to a doll house having a unique construction of an interlocking front wall and door assembly and an adjacent bay window. The doll house is of the collapsible type having a novel design in its construction. It is constructed of a small number of individual panels which comprise the walls, floors, roof, etc. and a bay window assembly. The panels are provided with grooves and slots so that all the panels slide together easily and support one another. No tools or screws are required for construction and the parts are locked tightly together in a rigid structure by the simple insertion of several small pegs in matching holes provided in the various panels. The doll house has various functional and decorative features including a chimney, rectangular decorative members and an address sign over the main doorway.

10 Claims, 11 Drawing Figures

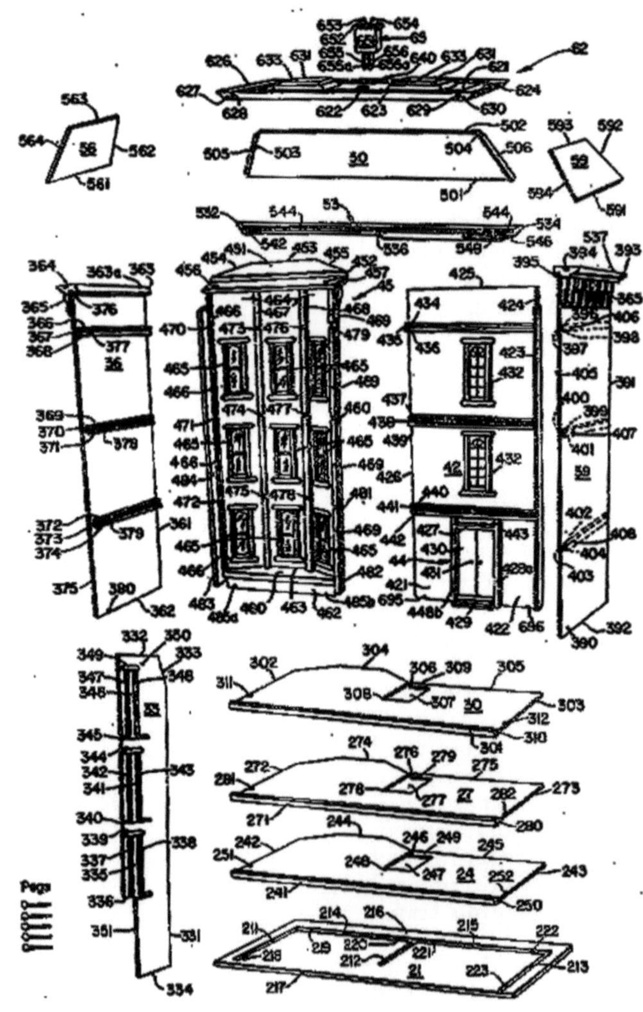

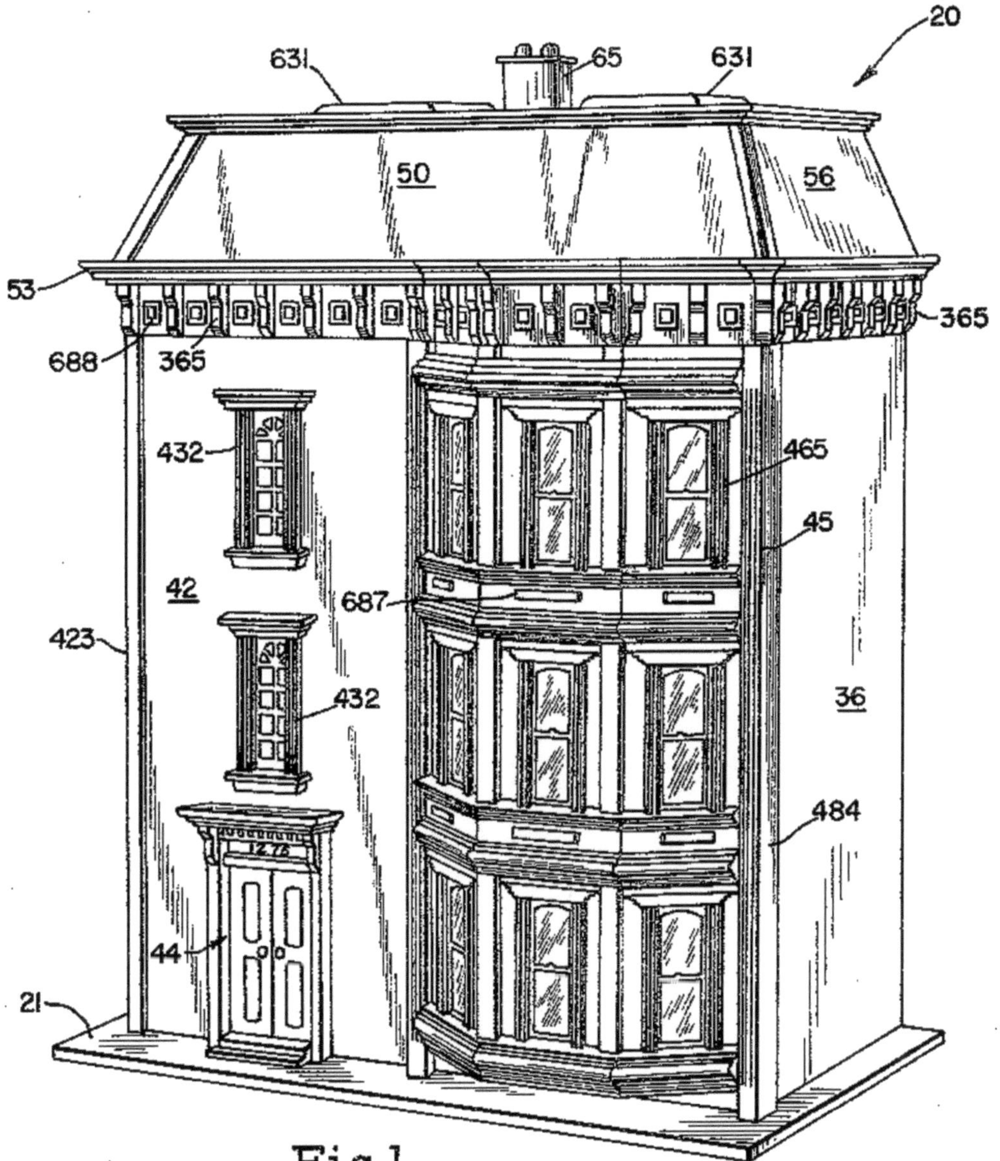

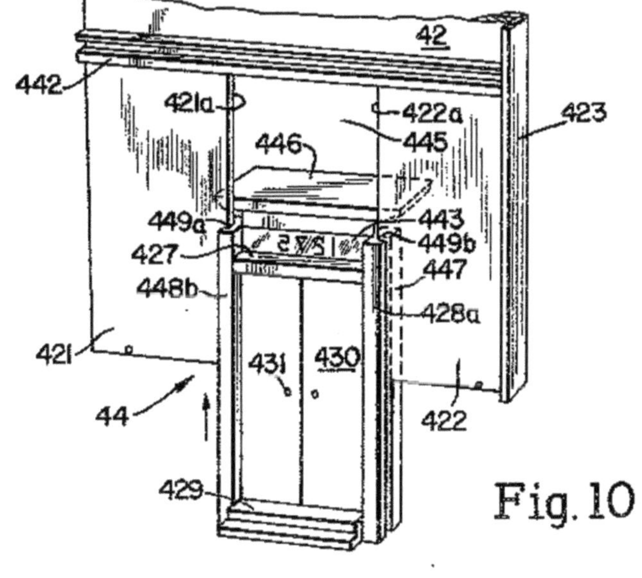

Fig.1

Fig.10

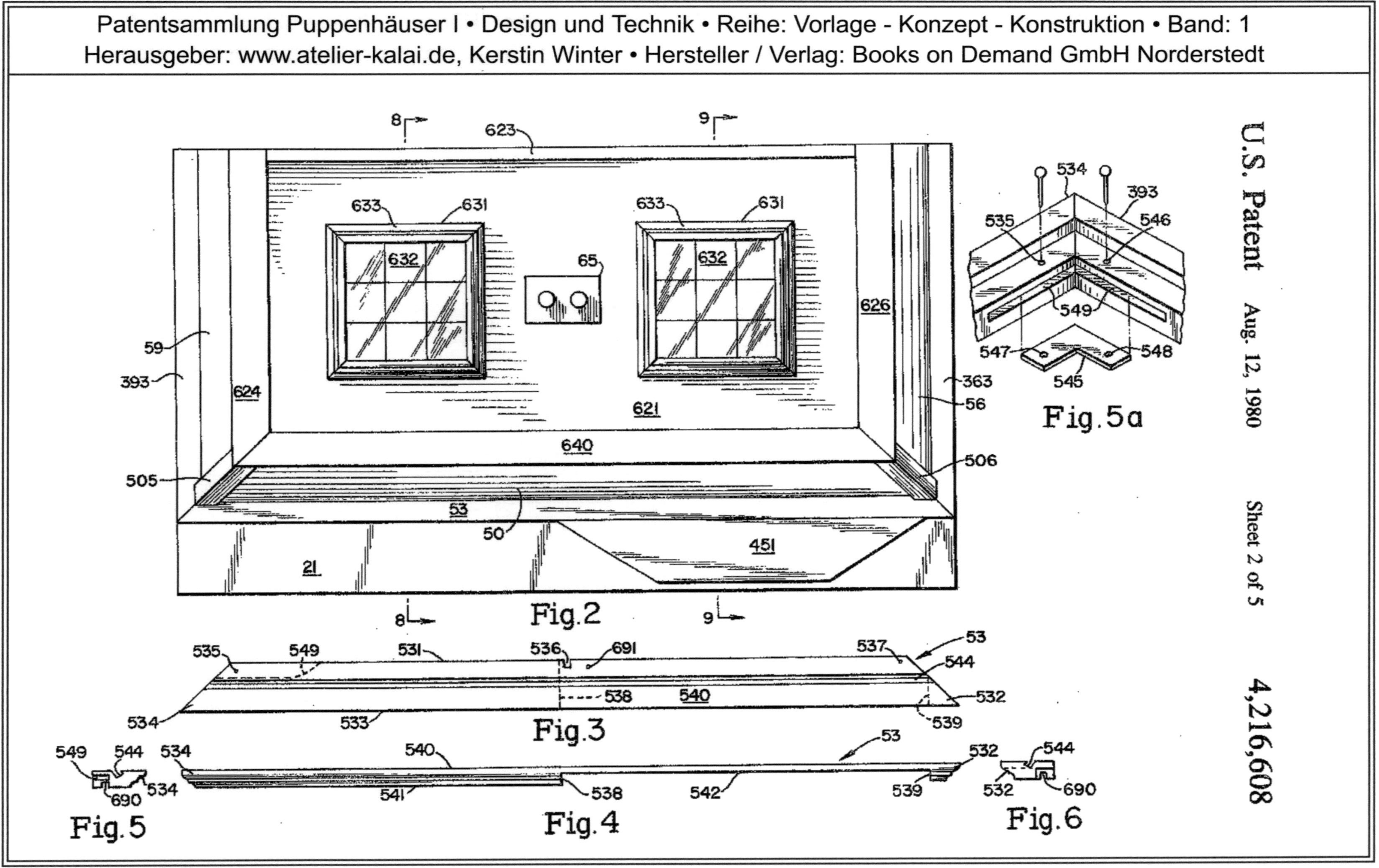

U.S. Patent
Aug. 12, 1980
Sheet 2 of 5
4,216,608
Fig.2
Fig.3
Fig.4
Fig.5
Fig.6
Fig.5a

Fig.7

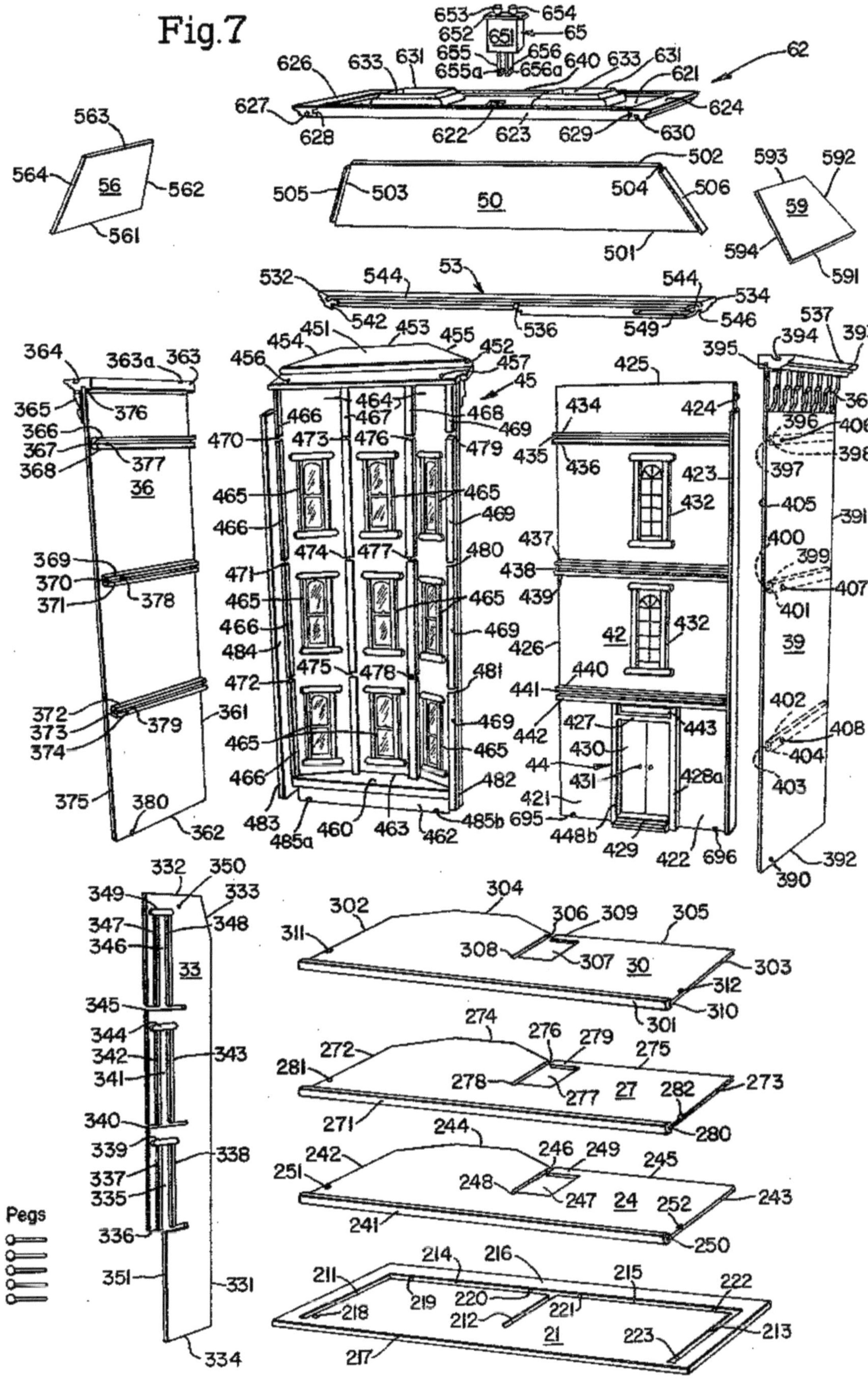

Fig. 8

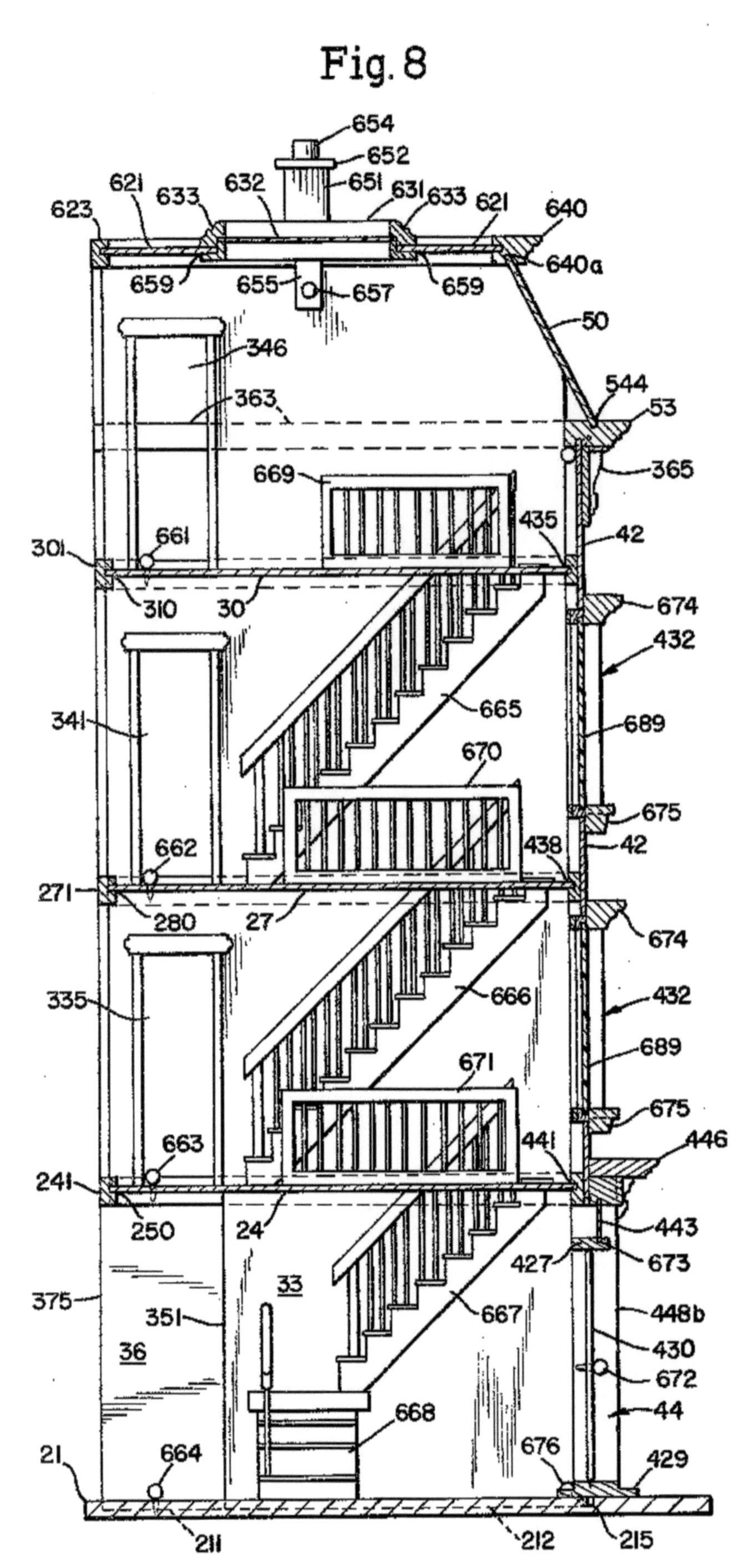

Fig. 9

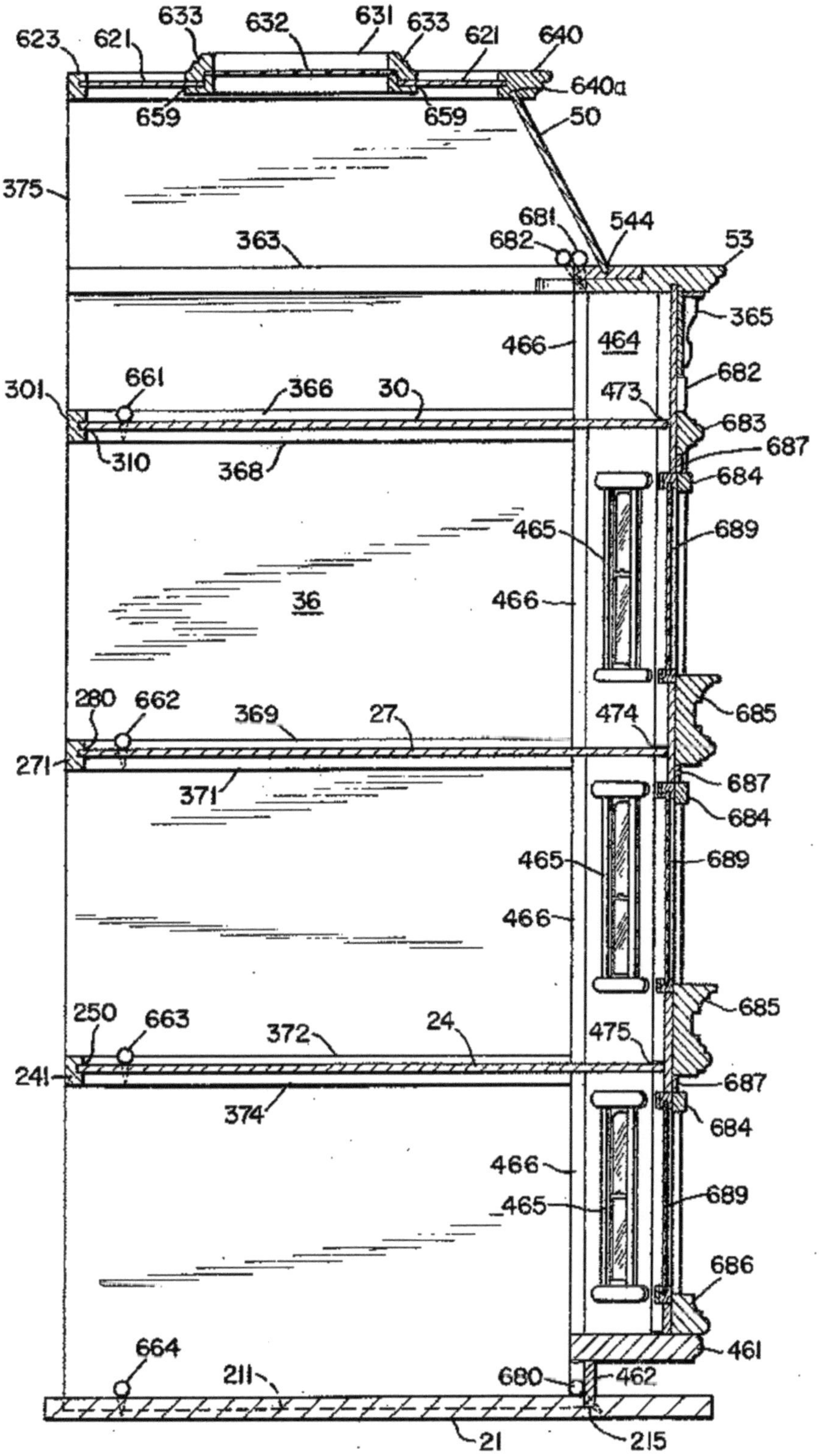

DOLL HOUSE

INTRODUCTION

This invention relates to a doll house of the knock-down or collapsible type of simplified construction capable of being easily constructed or taken apart.

Although there are many types of doll houses which provide recreation and education to children and adults, many of them are of the permanent type and present problems with shipping and storage. Others, which are of the knock-down type, are relatively complicated in construction and require various types of fasteners and locking members to hold the doll house together. These doll houses, for the most part require the use of tools and screws for their construction and lack the rigidity that is desired in a doll house when constructed. Furthermore, more sturdy types of doll house, which can be easily knocked-down, are desired by various adult hobbyists. These doll houses are required to have open access to the various floors in order that the hobbyists can set up various furniture arrangements, etc. Interior decorators also find the doll houses useful in planning the furniture arrangements, etc., of rooms. In particular, it has been desirable to have a doll house with a large number of rooms to provide for a large number of furniture arrangements, etc. The unique design of the present doll house provides a doll house which includes a plurality of rooms and provides a novel bay window construction for the doll house. In addition, the doll house includes a novel construction whereby the entranceway is attached to the doll house by a slide and groove means.

Applicant has developed a unique series of doll houses capable of being knocked-down as represented by U.S. Pat. Nos. 3,906,659; 3,996,693; 4,021,960; 4,018,001 and 4,094,090.

BRIEF SUMMARY OF THE INVENTION

This invention relates to a doll house of the knock-down or collapsible type of simplified construction, capable of being easily constructed or taken apart for convenient storage. A unique design provides for a front entrance which interlocks with the front wall of the doll house, a novel bay window construction, and a large number of rooms from a minimum number of extended floor and wall panels provided with grooves and slots so that all parts slide together easily and support one another. The assembling of the doll house is easily done in minutes and does not require the use of tools or screws. Once assembled, all the parts are locked together by the insertion of several small pegs between adjacent parts to provide a rigid structure.

It is, therefore, an object of this invention to provide a doll house having a unique construction design capable of being easily assembled or constructed without special skill or the use of various tools and fasteners, which can be easily disassembled or knocked-down for storage or transport purposes.

Another object of this invention is to provide a doll house constructed from a novel arrangement of floor and wall panels provided with grooves and slots whereby assembly or disassembly of the doll house is facilitated.

Another object of this invention is to provide a novel floor construction for a doll house whereby additional rooms are provided for the doll house.

Another object of this invention is to provide a doll house that is relatively rigid and sturdy in construction when assembled requiring only simple pegs for holding the assembled house together.

Another object of this invention is to provide a novel front door construction in the front wall of the doll house.

A further object of this invention is to provide a novel bay window construction in the front wall of the doll house.

Other objects, advantages and features of the invention will become apparent from the following detailed description of a preferred embodiment of the invention when considered with the accompanying drawings.

BRIEF DESCRIPTION OF THE DRAWINGS

FIG. 1 is a perspective front view of the doll house as fully assembled embodying the novel aspects of the invention;

FIG. 2 is a perspective top view of the doll house showing the novel features of the roof;

FIG. 3 is a perspective top view of the continuous front molding of the doll house;

FIG. 4 is a front edge view of the continuous front molding of FIG. 3;

FIG. 5 is a right end view of the continuous front molding of FIG. 3;

FIG. 5a is a perspective inner view of the right corner of the base of the roof showing the novel securing means thereof;

FIG. 6 is a left end view of the continuous front molding of FIG. 3.

FIG. 7 is an exploded perspective fragmentary view as seen from the rear, disclosing front, left and right side walls as well as the floors, roof, bay window and various parts of the doll house;

FIG. 8 is a cross-sectional view taken along line 8—8 of FIG. 2;

FIG. 9 is a cross-sectional view taken along line 9—9 of FIG. 2; and

FIG. 10 is an enlarged perspective rear view of the door and interlocking front wall.

DETAILED DESCRIPTION OF THE DRAWINGS

Referring to the drawings, FIG. 1 shows a perspective generally front view of a doll house embodying the principles of the invention. FIG. 1 shows the doll house fully assembled and ready for use as a recreational device or as a model house for display purposes with furniture arrangements, etc. FIG. 2 is a perspective top view of the doll house of FIG. 1. FIG. 7 is an exploded perspective rear view disclosing the various parts of the doll house and the manner in which the parts are assembled together.

The doll house comprises a rectangular base or first floor member 21 having grooves 211, 212, 213, 214 and 215 in the top surface, and peg holes 218, 219, 220, 221, 222 and 223. A left side wall member 36 comprises between edge 362, front edge 361, rear edge 375, an integral second floor horizontal beam member 374 containing a groove 373 and peg hole 379, an integral third floor horizontal beam member 371 containing a groove 370 and peg hole 378, and an integral fourth floor horizontal beam member 368 containing a groove 367 and peg hole 377. The top portion of the left wall comprises left molding member 363 with groove 376 and slanted groove 364. Similarly, a right side wall member 39

Patentsammlung Puppenhäuser I • Design und Technik • Reihe: Vorlage - Konzept - Konstruktion • Band: 1 Herausgeber: www.atelier-kalai.de, Kerstin Winter • Hersteller / Verlag: Books on Demand GmbH Norderstedt

3

comprises bottom edge 392, front edge 391, rear edge 405, integral second floor horizontal beam member 404 with groove 403 and peg hole 408, integral third floor horizontal beam member 401 with groove 400 and peg hole 407, integral fourth floor horizontal beam member 398 with groove 397 and peg holes 406 and right molding member 393 with groove 395 and slanted groove 394.

A front wall member 42 comprises top edge 425, bottom edge 442, right side edge 424, left side edge 426, integral L-shaped right corner beam 423, inner front wall edges 421a and 422a, an integral second floor horizontal beam member 440 containing a groove 441, an integral third floor horizontal beam member 437 containing a groove 438, an integral fourth horizontal beam member 434 containing a groove 435 and peg holes 695 and 696. Windows 432 and doorway assembly 44 are also disposed in the front wall member.

The doorway assembly 44, shown in FIG. 10, comprises door side members 428a and 448b, step member 429, door members 430, door overhang member 446 and cross member 427. Doorway assembly 44 attaches to front wall member 42 by sliding inner front wall edges 421a and 422a into inner grooves 449a and 449b, respectively, of door side members 428a and 448b, as shown in FIG. 10.

A front bay window member 45 comprises top bay member 451, with peg holes 456a, 456b and 457, right edge 482, left corner beam 484 with left edge 483, bay front wall members 464, integral vertical segmented beam members 466, 467, 468 and 469, bottom bay member 460 with vertical support member 462 and peg holes 485a and 485b. Windows 465 are disposed in bay front wall members 464. Integral vertical segmented beam members 466, 467, 468 and 469 are provided with spacings 470, 471, 472, 473, 474, 475, 476, 477, 478, 479, 480 and 481, respectively, for fitting the doll house securely together by interaction with the floor members.

An intermediate wall member 33 comprises front edge 331, rear edge 351, bottom edge 334, top edge 332, slanted front edge 333, horizontal floor slots 336, 340, 345, doors 335, 341, and 346, and peg hole 350. Door 335 comprises door side member 337 and 338 and top member 339, door 341 comprises door side members 342 and 343 and top member 344 and door 346 comprises door side members 347 and 348 and top member 349.

A second floor member 24 comprises front edge 245, front bay edge 244, rear edge molding member 241, left and right side edges 242 and 243, stair well opening 247 having open end slot 246 and slot 248, landing 249 and peg holes 251 and 252.

A third floor member 27 comprises front edge 275, front bay edge 274, rear edge molding member 271, left and right side edges 272 and 273, stair well 277 having open end slot 276 and slot 278, landing 279 and peg holes 281 and 282.

An attic or fourth floor member 30 comprises a front edge 305, front bay edge 304, rear edge molding member 301, left and right side edges 302 and 303, stair well 307 having open end slot 306 and slot 308, landing 309 and peg holes 311 and 312.

A front roof member 50 of the house comprises top edge 502, bottom edge 501 and left and right L-shaped members 505 and 506 on left and right edges 503 and 504, respectively.

A left side roof member 56 of the house comprises top edge 563, bottom edge 561, rear edge 564 and slanted

4

front edge 562. A right side roof member 59 comprises top edge 593, bottom edge 591, rear edge 594 and slanted front edge 592.

A top roof member 62, shown in detail in FIGS. 2 and 7, comprises planar member 621, front edge molding member 640, rear edge molding member 623, left and right side edge molding members 626 and 624 with slanted grooves 627 and 630 and grooves 628 and 629, chimney opening 622, and skylights 621 with edge moldings 633 and translucent panes 632.

A chimney 65 comprises structure 651, which comprises two downwardly projecting parallel members 655 and 656 having holes, 655a 656a respectively, and top member 652 with stacks 653 and 654.

A continuous front molding member 53, shown in detail in FIGS. 3 and 4, comprises left and right slanted front side edges 532 and 534, front edge 533, rear edge 531, rear center slot 536, slanted top groove 544, cut out bottom portion 542 and peg holes 535, 691 and 537. FIG. 5a shows the junction at the right front corner of front molding member 53 and right molding member 393 with inner combined L-shaped groove 549 wherein L-shaped locking member 545 is secured by pegs when the doll house is assembled. FIGS. 5 and 6 show left and right end views of continuous front molding member 53 of FIG. 4 showing bottom groove 690 and slanted top groove 544.

As shown in FIG. 1, a series of decorative rectangular members 688 are placed on the exterior of the house below the front and side moldings, said decorative rectangular members being disposed between decorative member 365.

The various parts of the doll house are adapted to be assembled together and held together with pegs. The assembly of the doll house is initiated by placing the wall members in the grooves of first floor member 21. Before doing so the doorway assembly 44 is introduced in the front wall by sliding the doorway assembly into the front wall. This is accomplished by sliding inner front wall edges 421a and 422a into grooves of door side members 428b and 428a, respectively. The bottom edge 422 of front wall member 42 is then inserted in groove 215 of the first floor member 21 and vertical support member 462 of bay window member 45 is inserted in groove 214 of the first floor 21. Bottom edges 362 and 392 of side walls 36 and 39 are inserted in grooves 211 and 213, respectively. The front edges 361 and 391 of walls 36 and 39 fit within left corner beam 484 and right corner beam 423, respectively. The three walls are retained in place by inserting pegs into peg holes 380 and 390 of side walls 36 and 39 and peg hole 218 and 223 of floor 21 and into peg holes 695 and 696 of front wall 42 and peg holes 221 and 222 of floor 21. Bay window member 45 is retained in place by inserting pegs into peg holes 485a and 485b of vertical support member 462 and into peg holes 219 and 220 of floor 21.

The three walls and bay window are further retained by placing continuous molding member 53 across and abutting the top of front wall member 42 and bay window member 45 with left edge 532 abutting left molding member 363, right edge 534 abutting right molding member 393 and cut out bottom portion 542 abutting top bay member 451. Continuous molding member 53, the three wall members and bay window members are retained on the left side by inserting pegs into aligned peg holes 537 and 456a, aligned peg holes 363a and 456b and aligned peg holes 691 and 457. The front wall member 42 and right side wall member 39 are retained at the

4,216,608

5

right front corner, as shown in FIG. 5a, wherein L-shaped locking member 545 is inserted into L-shaped slot 549, a peg is inserted into aligned peg holes 535 and 547 and a peg is inserted into aligned peg holes 546 and 548.

Intermediate wall 33 is next inserted from the rear of the house, in groove 212 of floor 21 and the front edge 331 is inserted in slot 536 of continuous molding member 53 at the junction of front wall member 42 and bay window member 45. Second floor member 24 is next assembled by sliding left and right side edges 242 and 243 simultaneously in groove 373 of horizontal beam member 374 and groove 403 of horizontal beam member 404 of side walls 36 and 39, respectively. The second floor member is pushed forward until its front edge 245 abuts against front wall 42 and front edge 244 abuts against bay window 45 and, in so doing the second floor is positioned with front edge portion 245 in groove 441, front edge portion 244 in spacings 472, 475, 478 and 481, and slots 246 and 248 of second floor member 24 are engaged with slot 336 of intermediate wall 33. The second floor member 24 and intermediate wall 33 are retained in place by inserting pegs in hole 251 of the second floor and hole 379 in groove 373 of horizontal beam member 374 on left side wall 36 (see FIGS. 8 and 9), and, similarly, a peg (not shown) is inserted in hole 252 of the second floor and hole 408 in groove 403 of horizontal beam member 404 on right side wall 39.

The third floor member 27 is next assembled by sliding left and right side edges 272 and 273 simultaneously in groove 370 of horizontal beam member 371 and groove 400 of horizontal beam member 401 of side walls 36 and 39, respectively. The third floor member is pushed forward until its front edge portion 275 abuts against front wall 42 and front edge portion 274 abuts against bay window 45 and, in so doing the third floor is positioned with front edge portion 275 in groove 438, front edge portion 274 in spacings 471, 474, 477 and 480 and slots 276 and 278 of third floor member 27 are engaged with slot 340 of intermediate wall 33. The third floor member 27 and intermediate wall are retained in place by inserting pegs in hole 281 of the third floor and hole 378 in groove 370 of horizontal beam member 371 on left side wall 36 (see FIGS. 8 and 9), and similarly, a peg (not shown) is inserted in hole 282 of the third floor and hole 407 in groove 400 of horizontal beam member 401 on right side wall 39.

The attic or fourth floor member 30 is next assembled by sliding left and right side edges 302 and 303 simultaneously in groove 367 of horizontal side beam 368 of side wall 36 and groove 397 of horizontal side beam 398 of side wall 39, respectively. The fourth floor member is pushed forward until its front edge portion 305 abuts against front wall 42 and front edge portion 304 abuts against bay window 45 and, in so doing the fourth floor is positioned with front edge portion 305 in groove 435, front edge portion 304 in spacings 470, 473, 476 and 479 and slots 306 and 308 of fourth floor member 30 are engaged with slot 345 of intermediate wall 33. The fourth floor member 30 and intermediate wall are retained in place by inserting pegs in hole 311 of the fourth floor and hole 377 in groove 367 of horizontal beam member 368 on left side wall 36 (see FIGS. 8 and 9), and similarly, a peg (not shown) is inserted in hole 282 of the fourth floor and hole 406 in groove 397 of horizontal beam member 398 on right side wall 39.

The roof is assembled by inserting bottom edge 501 of front roof member 50 into slanted groove 544 of contin-

6

uous molding member 53. Similarly, the bottom edge 561 of side roof member 56 is inserted in slanted groove 364 of molding 363 on left side wall 36 and the bottom edge 591 of side roof member 59 is inserted in slanted groove 394 of molding 393 on right side wall 39. The roof assembly is completed by placing top roof member 62 on the front and side roof members wherein top edge 502 of front roof member 50 is inserted in slanted groove 640a of front edge molding member 640 (see FIGS. 8 and 9), top edge 563 of side roof member 56 is inserted in slanted groove 627 of side edge molding member 626 and top edge 593 of side roof member 59 is inserted in slanted groove 630 of side edge molding 624.

At this stage of assembly, the roof structure is provided with means for providing rigidity by the design of the chimney structure 651 which comprises two downwardly projecting parallel members 655 and 656 (FIG. 7) having holes, 655a and 656a respectively. Opening 622 of the roof is bisected by edge 332 of intermediate wall 33. Members 655 and 656 are spaced apart a distance a little greater than the thickness of the intermediate wall 33. Chimney 65 is assembled in the roof by installing members 655 and 656 through opening 622 whereby members 655 and 656 straddle intermediate wall 33. Holes 655a and 656a are adapted to be aligned together and aligned with hole 350 of the intermediate wall when the chimney is assembled. A peg 657 (see FIGS. 8 and 9) is inserted through holes 655, 350 and 656, which provides rigidity and stability to the roof structure.

At this stage of assembly, the inner stairs are provided within the structure as shown in FIGS. 8 and 9. Stair case 668 is positioned to abut intermediate wall 33 and form a base for the bottom of stair case 667 when the top edge of stair case 667 is placed on landing 249 of second floor member 24. Stair case 666 is positioned with the bottom of stair case 666 abutting second floor member 24 behind stairwell 247 and the top of stair case 667 abutting third floor member 27 at landing 279. Stair case 665 is positioned with the bottom of stair case 665, abutting fourth floor member 30 and landing 309. Rails 671, 670 and 669 are placed on the second, third and fourth floors, respectively, adjacent stair wells 247, 277 and 307, respectively.

At this stage of assembly, door knobs (shown in FIG. 1) are inserted in holes 431 on door members 430. In addition, decorative rectangular members 688 may now be added.

This placement of the pegs for holding the doll house together is uniquely designed for easy assembly as well as providing the rigidity required for the doll house.

The doll house is knocked-down or disassembled by reversing the above procudure. The unique and novel design provides the benefits of knock-down construction. The use of grooves and slots provide means for easily sliding the various parts together, which support one another. The doll house is rigid in construction and all parts are held together tightly by the simple insertion of a relatively small number of strategically arranged small pegs. When disassembled, the parts of the doll house can be stacked together for easy storage or shipment.

Although the doll house of this invention has been disclosed heretofore as the preferred embodiment, wherein four rooms are available by using the one intermediate wall member 33, it is understood that the doll house can be constructed to contain more than one intermediate wall member thus providing eight or more

7

rooms. Furthermore, the number of windows used can be more or less than shown in the preferred embodiment.

The above description of the invention is deemed to be the most practical and efficient embodiment and it should be understood that the invention is not limited to such embodiments as heretofore indicated as there could be further changes made in the arrangements, parts without departing from the principle of the present invention within the scope of the accompanying claims.

What is claimed is:

1. An easily assembled knock-down doll house having a bay window assembly and interlocking door assembly, the parts of which are fitted together and held together only with pegs comprising:

a. a generally rectangular first floor member comprising left, right and front grooves in the top surface thereof, at least one intermediate groove in said surface parallel to said left and right grooves, and a series of in-line holes disposed forward of said front groove;

b. a front wall member having a size and shape to have its bottom edge engage said front grooves of said first floor member, comprising a right vertical L-shaped side beam, at least one horizontal beam member disposed on the inner side, a cut out portion in the portion of the front wall adjacent said front grooves of said first floor member at least two holes disposed at the bottom edge and inner front wall edges disposed in the front wall member;

c. a bay window assembly having a size and shape to have its bottom edge engage said front grooves of said first floor member, comprising three non-planar bay front wall members, integral vertical segmented beam members, horizontal top and bottom bay members, a vertical bottom bay member, a left vertical L-shaped beam member and at least one window in each bay front wall member;

d. left and right side walls having respective sizes and shapes to have their front edges engage said L-shaped beams of said front wall member and bay window member and their bottom edges engage said respective left and right grooves of said first floor, each of said side walls containing second, third and fourth inner horizontal beam members having inward side grooves, molding at the top edges and a groove disposed in the molding at the terminus of said top edges;

e. at least one intermediate wall member adapted to have its lower edge engage a respective said intermediate groove in said first floor and a said vertical slot between said front wall member and bay window member, comprising at least one intermediate horizontal slot extending forwardly from the rear edge;

f. a second floor member adapted to have its side edges engage said grooves of said second inner beam members of said side walls and comprising a stairwell near the front edge and at least one horizontal slot extending rearwardly from the front edge, said slot adapted to engage said horizontal slot of said intermediate wall member;

g. a third floor member adapted to have its side edges engage said grooves of said third inner beam members of said side walls and comprising a stairwell near the front edge and at least one horizontal slot extending rearwardly from the front edge, said slot

8

adapted to engage said horizontal slot of said intermediate wall member;

h. a fourth floor member adapted to have its side edges engage said grooves of said fourth inner beam members of said side walls and comprising a stairwell near the front edge and at least one horizontal slot extending rearwardly from the front edge, said slot adapted to engage said horizontal slot of said intermediate wall member;

i. front and left and right roof members disposed with bottom edges in a slanted groove in a molding along the base of the roof, said molding being secured to said front and side wall members, said top roof member having at least one skylight and a chimney;

j. a doorway assembly having side door members with inner grooves, step member, door members, door overhang member and cross member;

k. a continuous front molding adapted to abut and be secured to the front wall member, bay window member and side molding members of the left and right side walls; and

l. a plurality of pegs adapted to be inserted in holes contained in said beams, grooves, wall and floor members at their junctures with each other for retaining said assembled structure together.

2. The doll house of claim 1, wherein the inner grooves of the side door members of the doorway assembly interlock and are retained by the inner front wall edges.

3. The doll house of claim 1, wherein the front wall member is secured to at least one side wall member by use of a L-shaped locking member inserted into an L-shaped slot within the continuous front molding and the molding of the right sidewall and secured therein by pegs placed within aligned holes.

4. The doll house of claim 1, wherein the top roof member comprises two skylights.

5. The doll house of claim 1, wherein the outer front and side walls and bay window surfaces are decorated with rectangular members.

6. The doll house of claim 1, wherein the bay window assembly has nine windows.

7. In an easily assembled knock-down type of doll house comprising front and side wall members, front, side and top roof members, intermediate wall and floor members, the improvement which comprises:

a bay window member including front wall members and integral vertical segmented beam members, said wall and beam members forming a non-planar bay window member with horizontal bay members disposed at the top and bottom of said wall and beam members, a vertical bottom bay member disposed in the first floor member in a groove in said first floor member, said top horizontal bay member disposed adjacent to and retained by a continuous molding member and having said top and bottom bay members retained by pegs disposed within said top and bottom bay members and front wall, side wall and floor members.

8. In an easily assembled knock-down type of doll house comprising front and side wall members, front, side and top roof members, intermediate wall and floor members, the improvement which comprises:

a doorway assembly including side door members with inner grooves disposed therein, stop member, door members, door overhang member and cross member, wherein inner front wall edges are dis-

9

posed in side inner grooves of said side door member whereby said doorway assembly is secured by and within the front wall member.

9. In an easily assembled knock-down type of doll house comprising front and side wall members, intermediate wall and floor members, the improvement which comprises:

a continuous front molding adapted to abut and be secured to the front wall member, bay window member and side molding member of the left and right side walls.

10

10. In an easily assembled knock-down type of doll house comprising front and side members, intermediate wall and floor members, the improvement which comprises:

a means for securing adjacent wall members by use of a L-shaped locking member inserted into an L-shaped slot within the abutting corner moldings of the adjacent wall members; said L-shaped member secured by placing pegs in aligned holes within the moldings and L-shaped locking member.

* * * * *

J. B. CROSBY AND E. C. LOWE.
TOY BUILDING.
APPLICATION FILED NOV. 1, 1917.

1,316,690.

Patented Sept. 23, 1919.
6 SHEETS—SHEET 1.

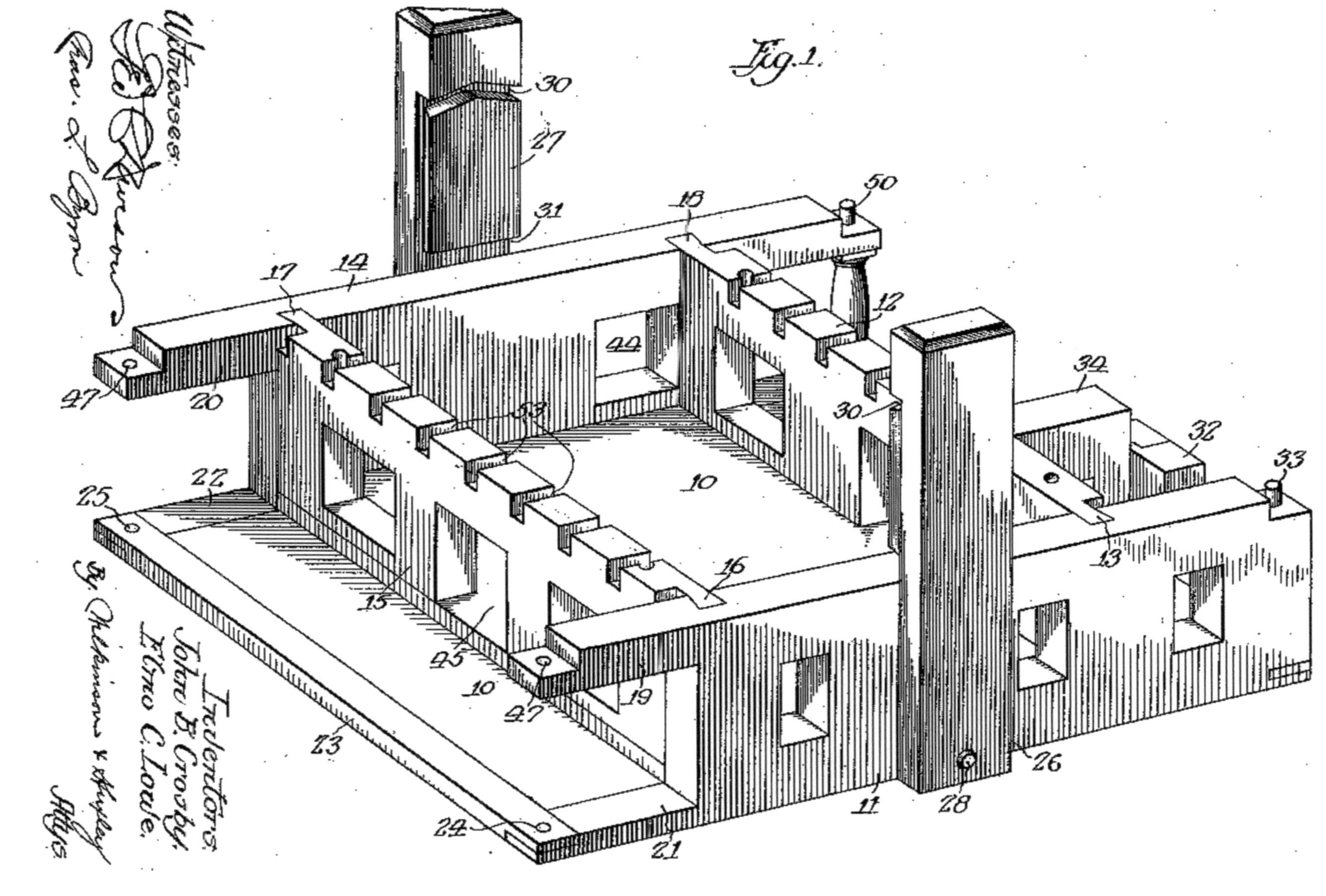

Witnesses.

Inventors.
John B. Crosby,
Elmo C. Lowe.
By Wilkinson & Huxley
Attys.

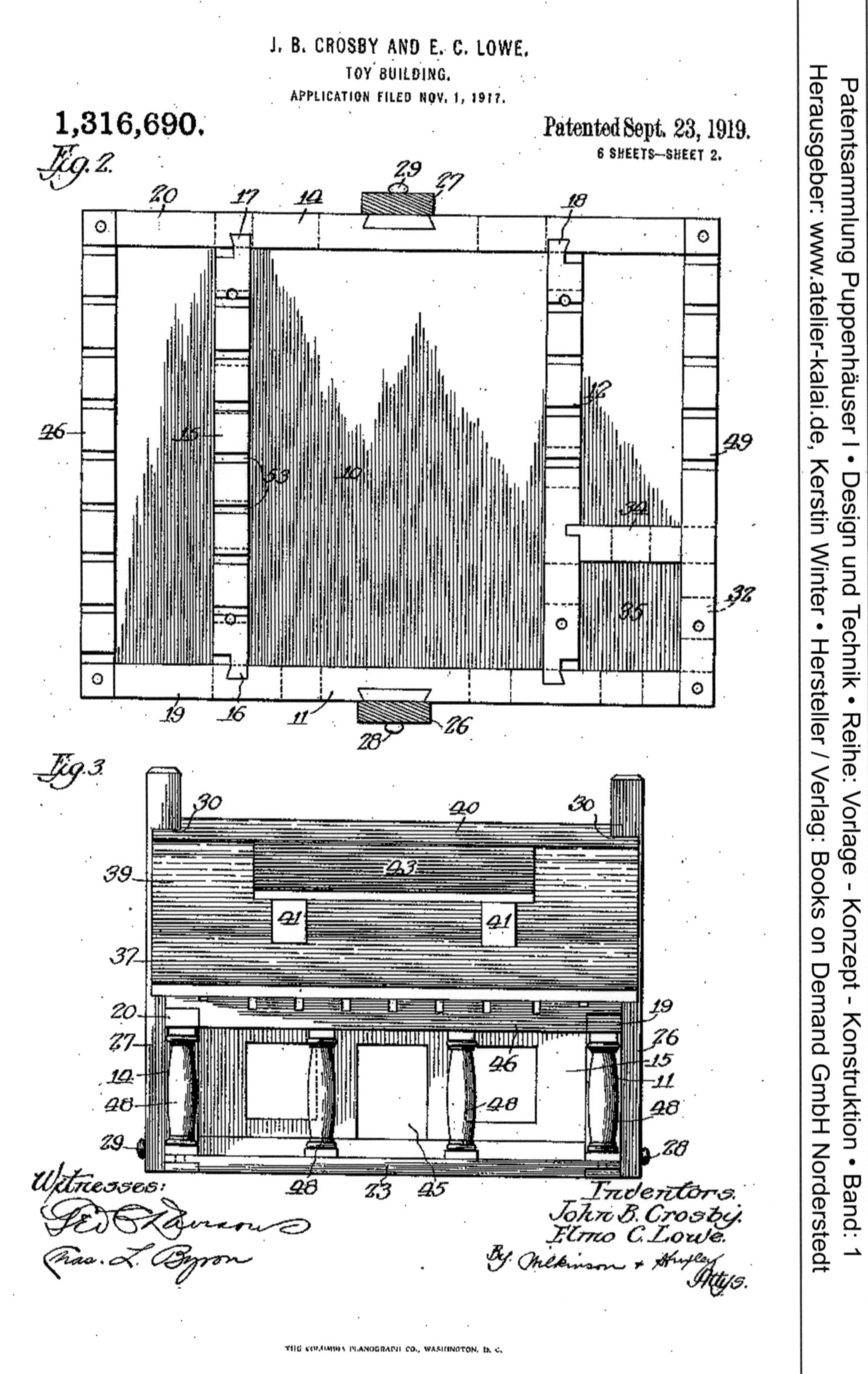

J. B. CROSBY AND E. C. LOWE.
TOY BUILDING.
APPLICATION FILED NOV. 1, 1917.
1,316,690.
Patented Sept. 23, 1919.
6 SHEETS—SHEET 2.
Fig.2.
Fig.3.
Witnesses:
Inventors.
John B. Crosby.
Elmo C. Lowe.
By Wilkinson + Huxley
Attys.
THE COLUMBIA PLANOGRAPH CO., WASHINGTON, D. C.

J. B. CROSBY AND E. C. LOWE.
TOY BUILDING.
APPLICATION FILED NOV. 1, 1917.

1,316,690.

Patented Sept. 23, 1919.

6 SHEETS—SHEET 3.

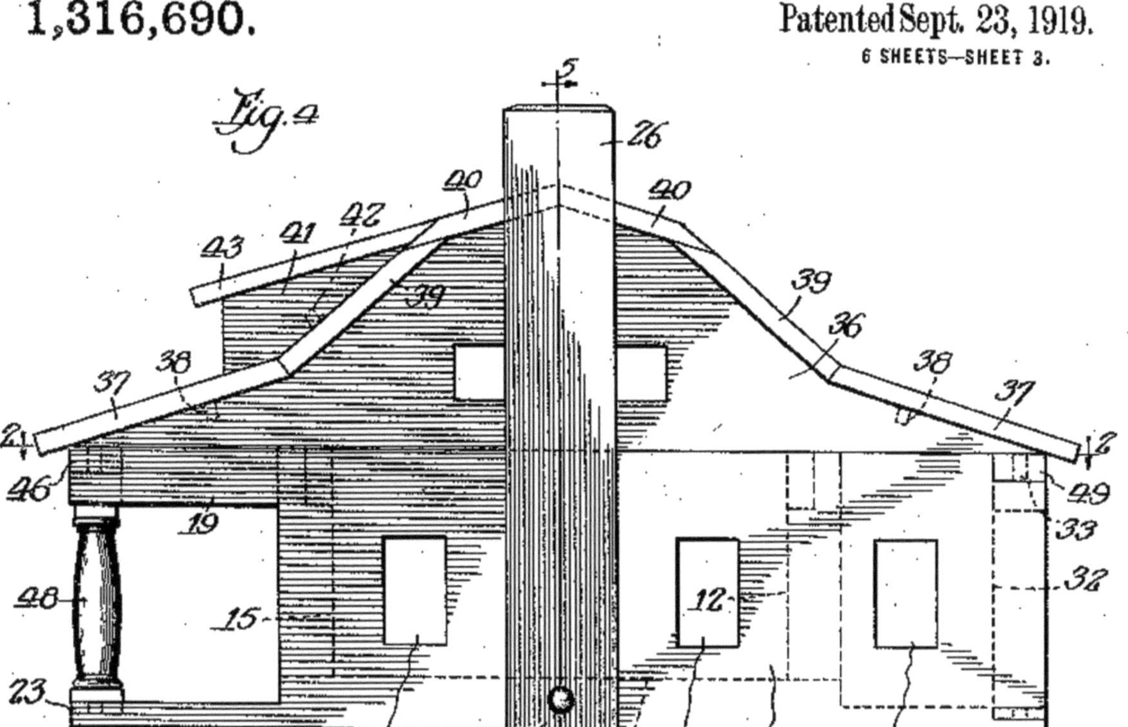

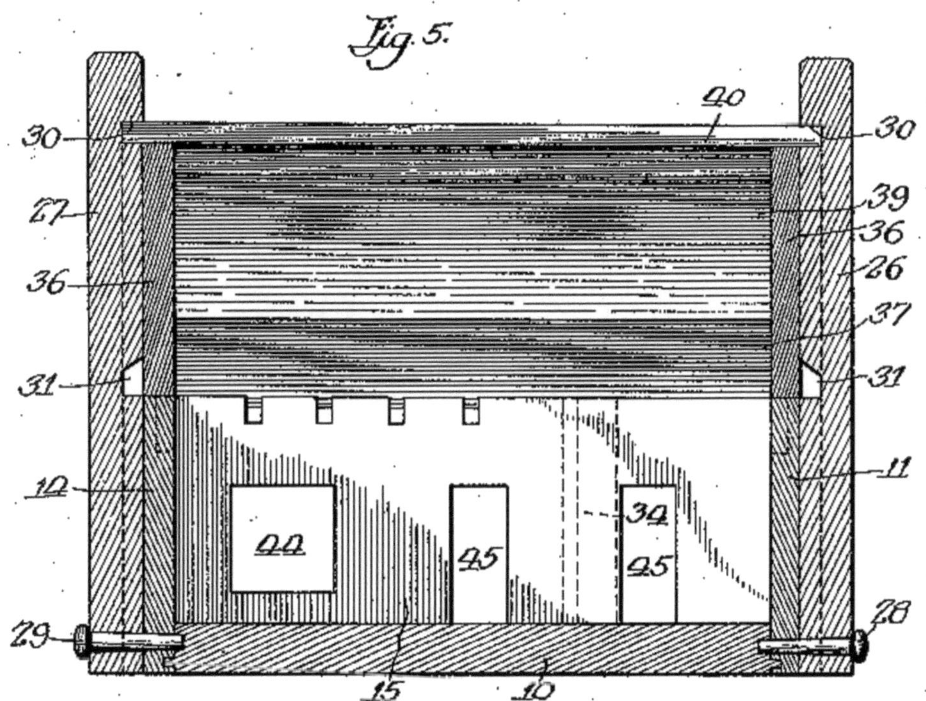

Witnesses:

Inventors.
John B. Crosby.
Elmo C. Lowe.

By Wilkinson & Huxley
Attys.

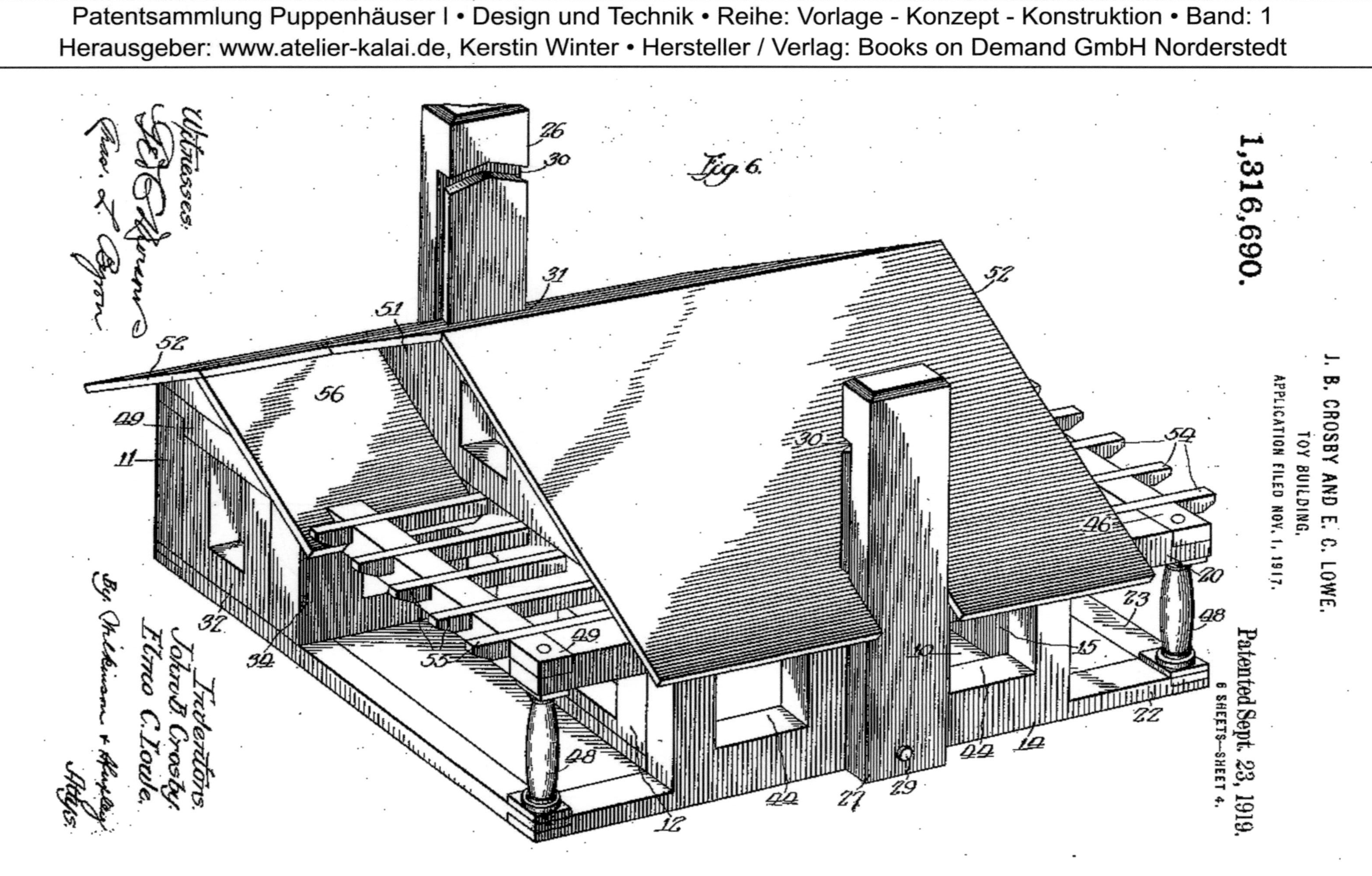
1,316,690.
J. B. CROSBY AND E. C. LOWE.
TOY BUILDING.
APPLICATION FILED NOV. 1, 1917.
Patented Sept. 23, 1919.
6 SHEETS—SHEET 4.
Fig. 6.
Witnesses:
Inventors:
John B. Crosby,
Elmo C. Lowe,
By Wilkinson & Huxley,
Attys.

1,316,690.

Patented Sept. 23, 1919.
6 SHEETS—SHEET 5.

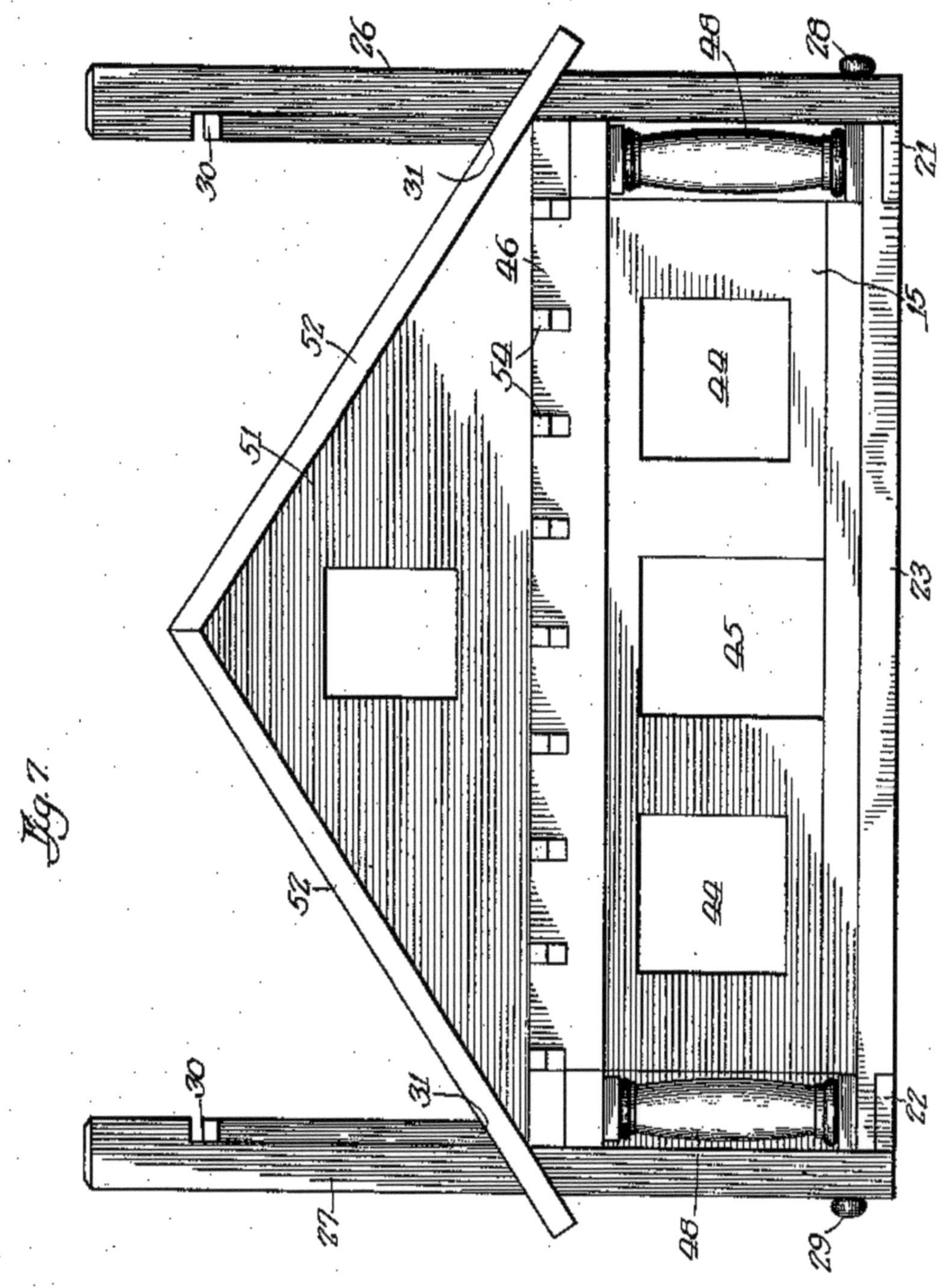

Fig. 7.

Witnesses:

Inventors.
John B. Crosby.
Elmo C. Lowe.
By. Wilkinson & Huxley
Attys.

Patentsammlung Puppenhäuser I • Design und Technik • Reihe: Vorlage - Konzept - Konstruktion • Band: 1
Herausgeber: www.atelier-kalai.de, Kerstin Winter • Hersteller / Verlag: Books on Demand GmbH Norderstedt

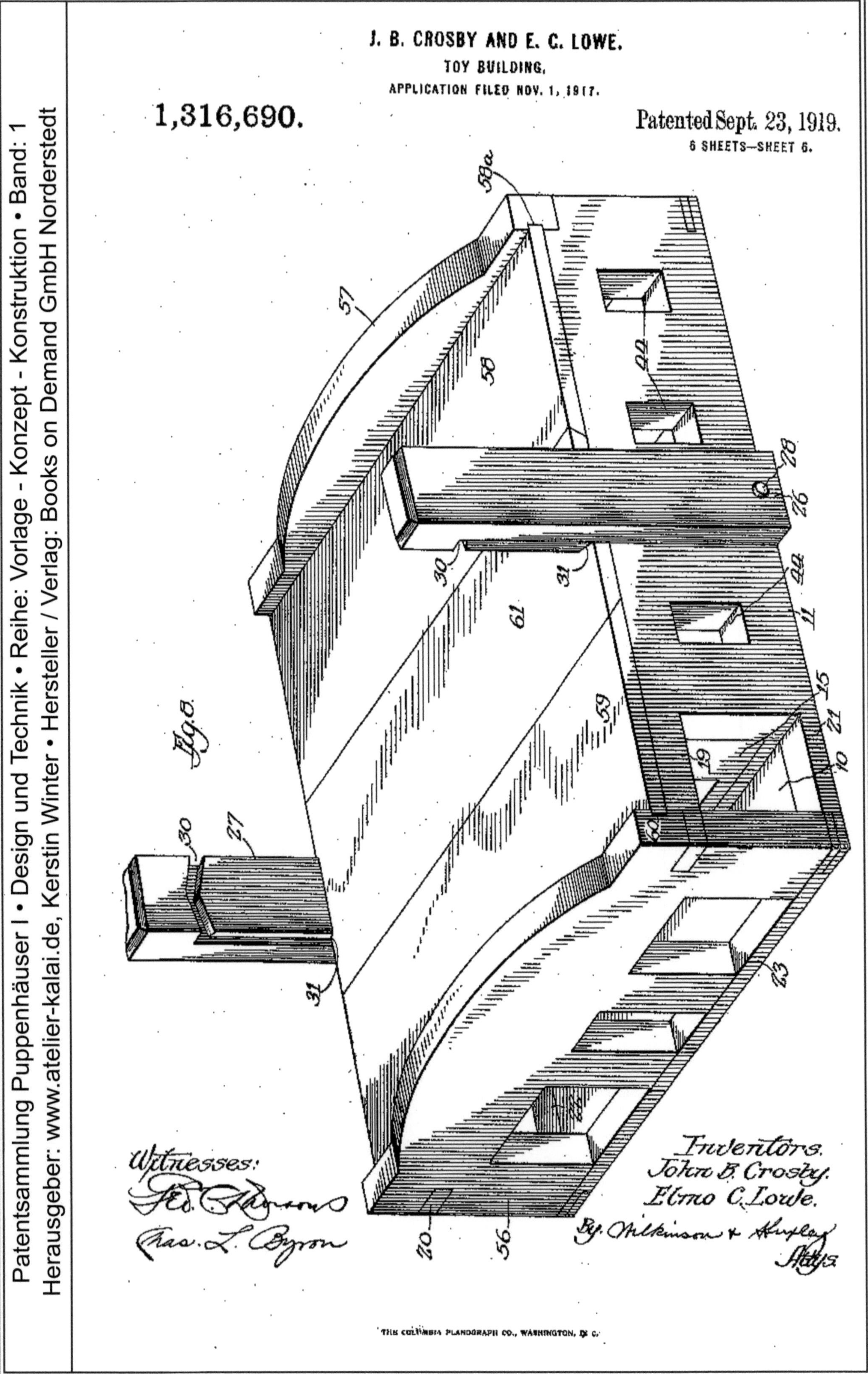

UNITED STATES PATENT OFFICE.

JOHN B. CROSBY, OF LAKE BLUFF, AND ELMO C. LOWE, OF EVANSTON, ILLINOIS.

TOY BUILDING.

1,316,690. Specification of Letters Patent. **Patented Sept. 23, 1919.**

Application filed November 1, 1917. Serial No. 199,632.

To all whom it may concern:

Be it known that we, JOHN B. CROSBY and ELMO C. LOWE, citizens of the United States, and residents, respectively, of Lake Bluff, in the county of Lake and State of Illinois, and of Evanston, in the county of Cook and State of Illinois, have invented certain new and useful Improvements in Toy Buildings, of which the following is a specification.

Our invention relates to toy buildings.

The main object is to construct, in a novel manner, buildings of different descriptions with the same blocks.

And another object is to provide a novel toy and at the same time an educational building equipment.

In our construction a main portion is erected upon which may be superimposed additional stories, together with a roof, all of which structure is then locked together by means of the chimneys. The structure thus erected will make either a different design of cottage or bungalow, or may be made into a store building, or the plan could be carried out further and any additional number of toy buildings could be erected.

Another advantage in our construction is that it is designed in correct architectural proportions so that a child is not only taught as to the different kinds of building construction, but also as to correct architectural proportions; thus the scheme serves as an educational as well as an amusement device.

And a further advantage in our toy building construction is that when the structure is complete it is locked together as a solid unit and may be picked up and carried about from place to place without danger of falling apart.

In the drawings we have shown three distinct types of building construction, the first representing a cottage, the second a bungalow and the third a store building.

Our invention will be further and better understood by reference to the accompanying drawings in which Figure 1 is a perspective view showing the main or base portion of the building;

Fig. 2 is a plan view of the main or base portion, parts being in section and taken along line 2—2 of Fig. 4;

Fig. 3 is a front elevation of a cottage;

Fig. 4 is a side elevation of the same cottage;

Fig. 5 is a sectional view taken in the plane of line 5—5 of Fig. 4;

Fig. 6 is a perspective view of a bungalow;

Fig. 7 is an elevation of the bungalow, and

Fig. 8 is a view in perspective showing a store construction.

Referring now specifically to the drawings in which like reference characters indicate like parts throughout, and particularly with reference to Fig. 1, 10 is the floor section of the main portion to which is attached by means of a grooved joint the vertical wall 11, also the vertical wall 12 which is joined to the wall 11 by means of a dove-tailed joint 13. 14 is a vertical wall having a groove at its bottom to which is attached the floor portion 10. 15 is a vertical side wall which rests upon the floor portion 10 and is joined at either end to the side walls 11, and 14. By means of four dove-tailed joints 13, 16, 17 and 18, the vertical members 14 and 11 are operatively joined and made rigid. Each of the members 11 and 14 have the top portions 19 and 20 extending outwardly to form supporting means for the roof of the porch. Portions 21, 22 extend outwardly to form bases for the porch section of the cottages and a portion of the floor for the store building. A transverse member 23 extends across from the lower portions 21 and 22 of each of the members 11 and 14 and is held by means of dowel pins inserted in the holes 24 and 25. Chimneys 26, 27 are attached by means of dovetailed joints to the members 11 and 14 and are held to the floor member 10 by means of wooden dowel pins 28, 29 inserted through the chimney and extending into the floor portions 10. These chimneys 26, 27 extend upwardly above the roof and form the locking means which holds the lower portion and the upper portion of the structure together. Transverse notches 30 in the chimney serve to hold the upper member of the roof in position when the cottage is complete and through notches 31, 31 in the chimneys, hold the roof members of the bungalow in position. At the rear end of the member 11 a short vertical member 32 is attached by means of a dowel pin 33 and extends transversely to another short vertical member 34 which is joined to the vertical member 12 at one end and to the member 32 at its other end, thus forming a separate chamber 35 at one corner of the structure.

Referring now particularly to Figs. 3, 4

and 5 which show the completed cottage, it will be seen that gable members 36—36 are set upon each of the vertical members 11, 14. Each of these gable members has a groove in its side adapted to slip over the dove-tailed portion of the chimney and is thus held in position in the structure. The roof members 37, 37 extend across from each of the gable members to the other and are held in position by means of the wooden pin 38. Two other roof members 39—39 extend across from each of the gable portions to the other and two other roof sections 40, 40 extend across the building on the tops of the gable sections and fit into the notches 30, 30 on each one of the chimneys. A gable member 41 having a dowel pin 42 covered by a roof member 43 forms a dormer window and may be fastened to the roof on either side. Window openings 44 may be located at any convenient place in the vertical walls as also may be door openings as at 45. Extending across from either of the projected portions 19 and 20 of the members 11 and 14 is a support 46 for the roof fastened to the portions 19 and 20 by means of dowel pins inserted in the holes 47, 47, the member 46 being supported and held in position by means of columns 48, 48, which may be of any shape or contour desired. Extending across from the rear portions of each of the members 11 and 14 is a member 49 similar to the transverse member 46 which forms a support for the lower roof member 37 and is held in position by means of the dowel pins 33 and 50.

It will thus be seen that by the addition to the main portion of the structure shown in Fig. 1, a complete cottage is erected.

Referring now specifically to Figs. 6 and 7 which illustrate a bungalow, we add to the structure shown in Fig. 1 right-angled gable members 51, 51 which rest upon the members 12 and 15. Roof members 52, 52 fit on either side of the gable members 51, 51 and into the transverse notches 31, 31 of the chimneys 26, 27 and are thus held securely in position, the chimneys remaining in the original position in which they are shown in Fig. 1 and fastened to the floor member 10 by means of the dowel pins 28 and 29. Resting in notches 53, 53 in the top of the member 15 and extending outward therefrom are rafters 54, 54 the outer ends of which rest in the notches in the transverse member 46. Resting in similar notches in the member 12 are also rafters 55, 55 which rest in notches cut in the top of the member 49. Small gable portions are adapted to fit on the walls which form the chamber 10 and are covered over by one end of one of the main roof portions 52 and by another roof portion 56. One of the columns 48 is inserted under the end portion 20 of the member 14 and serves to support this end portion 20 and is held in position by

means of a dowel pin inserted through the opening 47 in the portion 20 into a hole in the top of the column 48. A dowel pin is also inserted through the hole 25 of the members 22 and 23 into a hole in the bottom of the column which is thus held in place. Columns 48 may be positioned at any convenient place between the members 23 and 46.

Referring now specifically to Fig. 8, a front 56 having door and window openings therein, it fitted to the end portions 19 and 20 of the members 11 and 14. A member 57 extends across from the rear ends of the members 11 and 14 and forms the rear portion of the store. A flat roof portion 58 extending directly across from the top of each of the members 11 and 14 forms the rear portion of the roof, the member 57 having a recess 58ᵃ cut therein to receive one edge of the roof member 58. Extending across the front portion of the store is another roof member 59, the member 56 having a recess 60 therein which is adapted to receive one edge of the roof member 59. Another roof member 61, the ends of which fit into the recesses 31, 31 in the chimneys 26, 27, rest directly on the top of each of the members 11 and 14. The outside edges of the member 61 are beveled and fit snugly against the beveled edges of each of the members 58 and 59, and thus the whole roof is held securely in position when the member 61 is in proper position.

It will thus be seen that three separate and distinct types of buildings can be erected with one set of blocks thus affording a child considerable scope within which to exercise his idea of building, and the different architectural proportions in general.

For the store construction it would be possible of course to leave out each of the vertical walls 15 and 12 and still the structure would be complete and when each of the structures is complete it will be noted that they are locked securely together so that destructive force would be required to separate the parts and, in fact, the building may be turned upside down and yet be held securely together.

Many modifications will readily suggest themselves to persons skilled in the art of toy building, all of which we consider within the spirit and scope of our invention.

We claim:

1. In a toy building construction, the combination of a main portion, a secondary portion superimposed thereon, and a vertically extending chimney for securing said portions in locked relation with each other.

2. In a toy building construction, the combination of a main portion consisting of a floor member, vertical wall members secured to said floor member, one pair of said wall members secured at their ends to the other

wall members, a secondary portion consisting of gable members and roof members superimposed on said main portion, and a chimney adapted to secure said portions in locked relation with each other.

3. In a toy building, the combination of a floor member, vertical wall members secured to said floor members, one pair of said wall members secured at their ends to the other pair of vertical wall members and in spaced relation with each other, gable and roof members superimposed on said vertical wall members, and locking means for securing said wall members and said gable and roof members together.

4. In a toy building, the combination of a floor member, two pairs of vertical side members, one pair secured to said floor member, the other pair secured at their ends to the other vertical side member and spaced apart from each other, transverse bars secured to one pair of said side members at both ends of said members and supported by vertical columns, gable and roof portions superimposed upon said vertical walls, and chimneys secured to one pair of said side walls and to the bottom member, said chimneys having recesses therein to receive and hold said roof members.

5. In a toy building, the combination of a plurality of structural parts, and a chimney forming a locking means between said parts.

6. In a toy building, the combination of a body portion, a roof, and a chimney for locking the body portion and roof in a given relationship.

7. In a toy building, the combination of a main body portion, and a chimney whereby superstructures of different characters may be secured to said main body portion.

8. In a toy building, the combination of a main body portion, means whereby superstructures of different characters may be secured to said main body portion, and a chimney for locking together said main body portion and any of said superstructures.

9. In a toy building the combination of a main portion, comprising a floor and two pairs of side walls, a roof and gable portion superimposed upon said main portion, and chimney portions secured to one pair of said side walls for locking the structure together.

Signed at Chicago, Ill., this 25th day of October, 1917.

JOHN B. CROSBY.
ELMO C. LOWE.

[54] **DOLL HOUSE WITH CONNECTOR ELEMENT CONNECTING THREE WALL MEMBERS**

[76] Inventors: Harry E. Walmer, 721 N. Overlook Dr., Alexandria, Va. 22305; Judd Horbaly, 125 Commonwealth Ave., Alexandria, Va. 22309

[21] Appl. No.: **115,479**

[22] Filed: **Jan. 25, 1980**

[51] Int. Cl.³ A63H 3/52
[52] U.S. Cl. 46/19
[58] Field of Search 46/18, 19, 20, 21; 35/16

[56] **References Cited**

U.S. PATENT DOCUMENTS

1,337,171	4/1920	Ward	46/19
1,449,519	3/1923	Layton	35/16
2,441,761	5/1948	Guelicher	46/19
2,968,118	1/1961	Paulson	46/19
3,751,848	8/1973	Ahlstrand	46/19
3,906,659	9/1975	Walmer	46/19
4,058,909	11/1977	Poleri	35/16

Primary Examiner—F. Barry Shay
Attorney, Agent, or Firm—James C. Wray

[57]   **ABSTRACT**

A doll house has a base with a series of interconnected grooves, a first front wall, two tall vertical front walls, a shorter side wall and a shorter interior wall. A floor member engages those walls and cantilevers outward over the shorter interior wall. Together the floor member and base and walls form a rigid frame to which remaining panels of the doll house are connected. A front door assembly with side lights and platform steps are grooved to be tightly held within a door opening. A second short front wall and a side wall are captured between the cantilevered portion of the floor member and the base, further promoting stability. An intermediate floor and front, top, and side roof panels and chimney connectors complete the basic structure.

9 Claims, 10 Drawing Figures

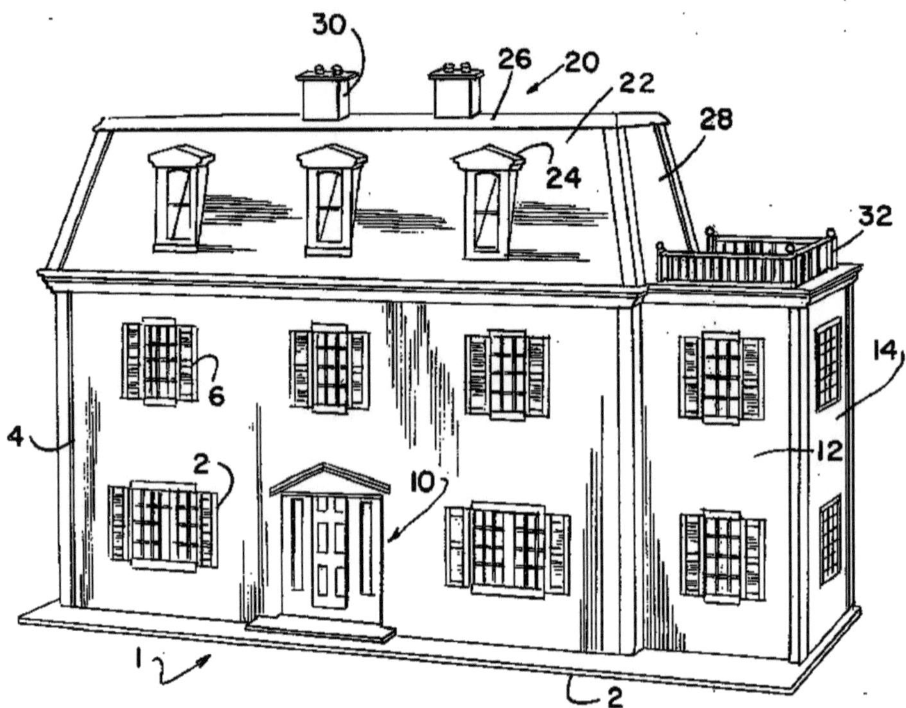

4,306,371

11

member directly above said downwardly disposed intermediate grooved beam member with an inwardly slanting groove disposed therein, a downwardly disposed front beam member perpendicular to and running between said side horizontal beam and downwardly disposed intermediate grooved beam member at the foremost edge of said floor member, and at least one horizontal slot extending rearwardly from the front edge, said slot adapted to engage an intermediate horizontal slot of said relatively tall inner wall member,

front, left and right roof wall members disposed with bottom edges in said slanted grooves, said front roof wall member having at least two dormer window units disposed therein,

a top roof member having a chimney hole disposed therein, left, right and front beam memers with outwardly slanted grooves disposed thereunder, a rear edge molding beam and said front beam mem-

12

ber having inwardly disposed notches aligned with said relatively tall inner wall member and said chimney hole, said outwardly slanting grooves in said front, left and right beam respectively engaging the top edges of said front, left and right roof wall members,

a chimney member containing a pair of parallel downwardly projecting members is received within said chimney hole with the downwardly projecting members stradling said relatively tall inner wall member and is secured thereto by a peg inserted in aligned holes in said relatively tall inner wall member and said parallel downwardly projecting members,

a plurality of pegs adapted to be inserted in said holes at their junctures with each other for retaining said assembled structure together.

* * * * *

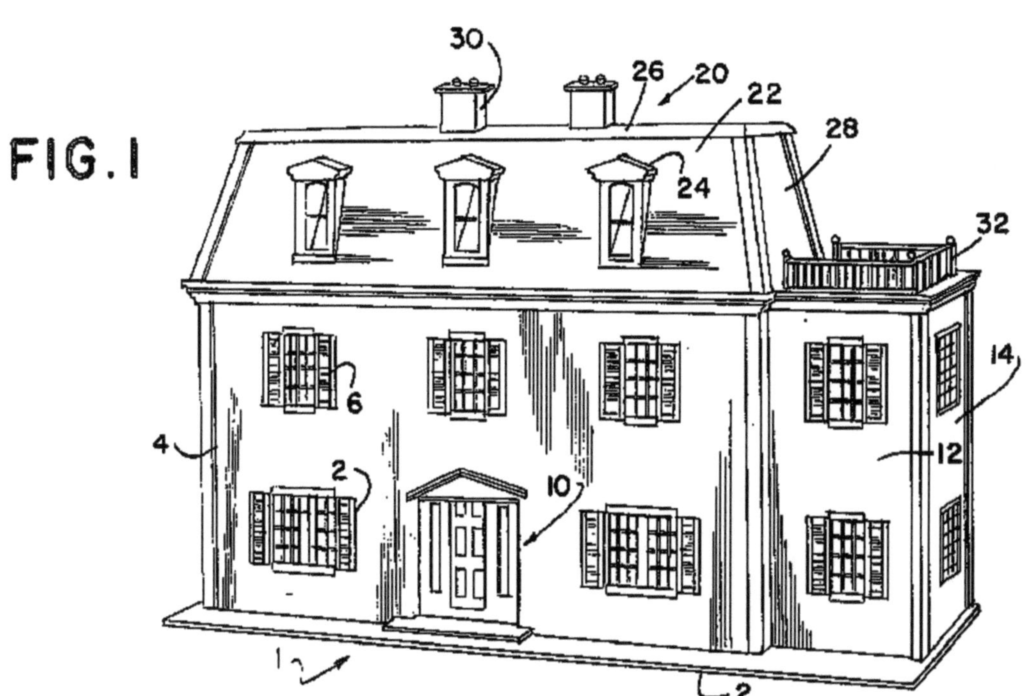

FIG.1
30
26
20
22
28
24
32
14
12
6
4
2
10

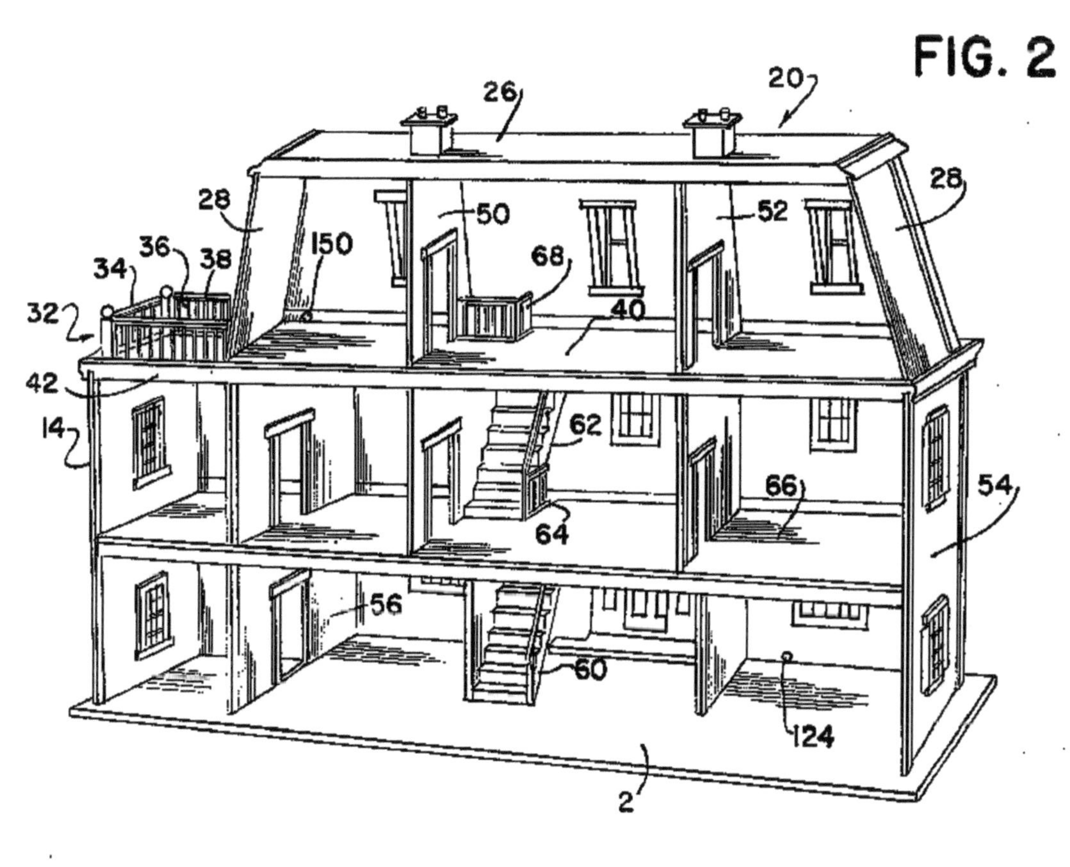

FIG. 2
26
20
28
50
52
34 36 38 150
68
32
42
40
14
62
66
54
64
56
60
124
2

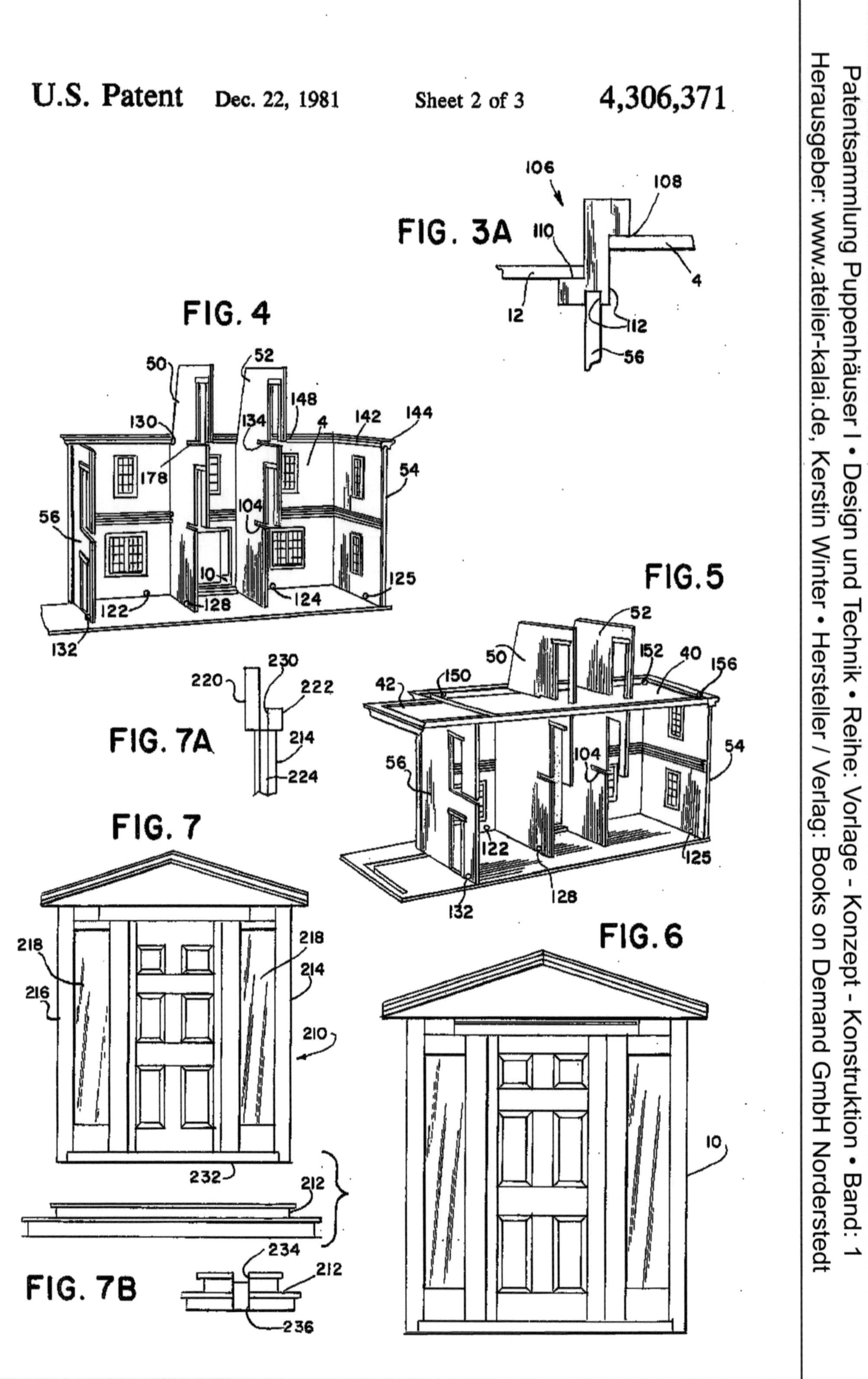
FIG. 3A
106
108
110
12
112
56
4
FIG. 4
50
52
148
130
134
4
142
144
178
54
56
104
10
125
122
128
124
132
FIG. 7A
220
230
222
214
224
FIG. 7
218
218
216
214
210
232
212
234
212
236
FIG. 5
50
52
152
40
156
42
150
56
104
54
122
128
125
132
FIG. 6
10

204
30
186 190 182 24
188 206
208 184 22 202
196 28
180 194
162 92
94 94 192 90
192 4
198
228 96
98
104
226 54
12 160 56 50 52 4 86
164 146
42 200 40
170 120
14 168 140
178 102 174 172
100
82 72 70
84 80 74 76 78
166 2

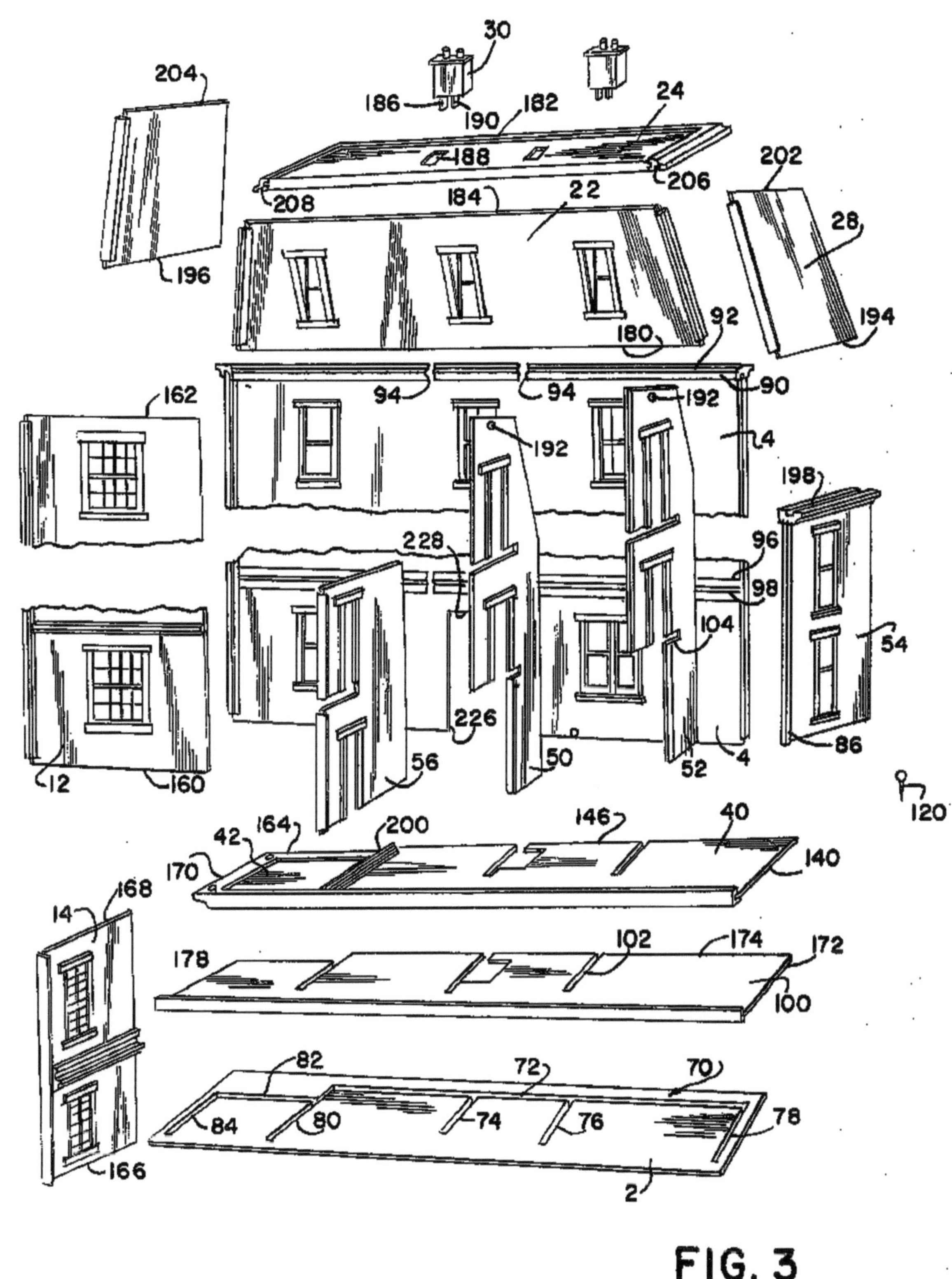

FIG. 3

DOLL HOUSE WITH CONNECTOR ELEMENT CONNECTING THREE WALL MEMBERS

BACKGROUND OF THE INVENTION

This invention relates to doll houses and toy houses particulary of the knock-down or collapsible type of simplified construction which are capable of being easily constructed or taken apart.

Historically many types of doll houses and toy houses have been provided. Most of the houses are permanently constructed, with attendant problems of shipping and storage. Knock-down houses of prior art design have required special tools and difficult assembly steps for construction. Once constructed, knock-down houses of prior art designs have experienced some difficulties in stability.

The inventors of the present invention have developed knock-down doll houses which go together simply with minimal or no tools and which remain stable once constructed. The present invention is a result of continued efforts to improve ease in assembly and rigidity of assembled construction of knock-down doll houses.

SUMMARY OF THE INVENTION

A doll house has a base with a series of interconnected grooves, a first front wall, two tall vertical front walls, a shorter side wall and a shorter interior wall. A floor member engages those walls and cantilevers outward over the shorter interior wall. Together the floor member and base and walls from a rigid frame to which remaining panels of the doll house are connected. A front door assembly with side lights and platform steps are grooved to be tightly held within a door opening. A second short front wall and a side wall are captured between the cantilevered portion of the floor member and the base, further promoting stability. An intermediate floor and front, top, and side roof panels and chimney connectors complete the basic structure.

The present invention provides rigidity in construction by cantilevering a main horizontal member and then placing offset vertical member below the cantilevered portion of the horizontal member. This ensures the rigidity of the basic structure and enables the removal of holding members required for intermediate steps in the construction.

The present invention increases the space within the knock-down doll house and increases the panel members which make up the doll house while at the same time enabling the reduction of non-aesthetic holding members which are less desirable in the house construction.

An easily aseembled knock-down doll house apparatus includes a base having plural connected angularly oriented slots for receiving bottom edges of vertical walls, the slots comprising a first elongated slot slightly spaced inward from a first elongated edge of the base for receiving a front vertical wall and plural slots communicating with the first slot and extending angularly therefrom toward a second opposite edge of the base, and a further slot extending angularly outward from an outer one of the plural slots and extending toward an edge of the base intermediate the first and second elongated edges, a remote slot extending from a portion of the further slot remote from the outer one of the plural slots toward the second edge of the base, a front vertical wall member mounted in the first slot, vertical trans-

verse wall members mounted in the plural slots, the vertical transverse wall members comprising at least one relatively short outer wall member and at least one relatively tall inner member, a floor member connected to the vertical transverse wall members parallel to the base and space therefrom, the floor member extending over the outer vertical transverse wall members and through the inner transverse wall members, and the floor member cantilevering outward beyond the vertical wall member positioned in the first outer slot and over the further slot and remote slot, and means for connecting the base, the vertical transverse wall members and the floor member for forming thereby a rigid skeletal frame of the base, vertical transverse wall members and floor member by which a doll house comprising the members is rigidly supported.

The doll house apparatus of the invention includes a second relatively short frontal wall member mounted in the further slot and extending between the base and a cantilevered portion of the floor member and having a first vertical edge connected to a relatively short outer wall member, a side wall member mounted in the remote slot and extending between the base and a cantilevered portion of the floor member and having a first vertical edge connected to a vertical edge of the second frontal member which is spaced from the relatively short outer wall member.

In apparatus of the invention the tall inner members comprise inward and upward sloping frontal edges at positions thereof above the floor member and the house includes a roof front member having a lower edge connected to a front edge of the floor member, the roof front member sloping inwardly and resting against the inward sloped edges of the relatively tall inner members, a roof top member resting on upper edges of the relatively tall inner members and having a front edge connected to an upper edge of the roof front member and means for connecting the roof top member to the relatively tall inner members thereby cooperating with the roof top member and the floor member to hold the roof front member in assembled condition, the floor member and the roof top member having complimentary upward and downward openings slot means respectively, and first and second roof side members mounted in the slot means.

The doll house apparatus of the invention has upward opening slots in the floor member are positioned immediately above the relatively short vertical wall members and wherein a portion of the floor member cantilevered outward beyond the slot comprises an outer deck member.

In apparatus of the doll house the vertical transverse wall members comprise horizontal slits extending partially through the wall members and further includes an intermediate floor member having a plan form similar to a plan from of the first said floor member and fitting within the slits and engaging in rearward opening slot means positioned on the frontal wall member.

Doll house apparatus includes a base having angularly oriented slot means for receiving vertical walls, plural vertical walls for inserting the grooves, at least one of the vertical walls having a door opening extending upward into the wall from a bottom edge of the wall, a door assembly for fitting in the opening, the door assembly comprising side members having outwardly facing first and second vertical slots for receiving opposite vertical edges of the door opening in the slots and

Patentsammlung Puppenhäuser I • Design und Technik • Reihe: Vorlage - Konzept - Konstruktion • Band: 1
Herausgeber: www.atelier-kalai.de, Kerstin Winter • Hersteller / Verlag: Books on Demand GmbH Norderstedt

3

the door assembly having an outer facing on one side of the slots and having an interior facing on the other side of the slots and a movable door hinged to one of the side members and abuting the other of said side members.

In apparatus of the invention the door assembly further includes a top member connected to the side members, the top member having an outer facial portion and an inner facial portion, and having an upward opening horizontal slot between the inner and outer facial portions for receiving an upper edge of the door opening.

The apparatus of the invention includes a step assembly, the step assembly having first and second platforms supported respectively on first and second riser blocks, the first and second platforms and first and second riser blocks having opposite outward opening vertical grooves for receiving outer edges of the door opening.

In easily assembled knock-down doll houses having an interlocking door assembly, the parts are fitted together and held together only with pegs.

A generally rectangular first floor member includes front and left and right side grooves in the top surface thereof, at least one intermediate groove in said surface parallel to said left and right grooves, a groove parallel, outside of and shorter in length than said left and right side grooves, a groove parallel to but not overlapping the length of said front groove and which runs perpendicular to and between the side groove adjacent the shorter outside groove and the foremost end of said shorter outside groove, and a series of in-line holes disposed behind said front groove.

A front wall member has a size and shape to have its bottom edge engage said front grooves of said first floor member, comprising left and right vertical L-shaped side beams, at least one segmented horizontal beam member disposed on the inner side, one top horizontal beam member containing a top inwardly slanted groove, an inward side groove, and which also has inwardly disposed vertical slots corresponding with the notches formed of the space between said segments of inwardly disposed horizontal beam member and the intermediate grooves in said first floor member, a cut out portion in the portion of the front wall adjacent said front grooves of said first floor member, at least two holes disposed at the bottom edge and inner front wall edges disposed in the front wall member.

A recessed front wall member has a size and shape to have its bottom edge engage said groove that is parallel to said front groove in said first floor member, a side edge to engage said vertical slotted beam in said intermediate side wall, and a top edge adapted to engage said downwardly disposed horizontal front beam member in said third floor member, and which comprises a vertical L-shaped side beam, and at least one horizontal beam member disposed on the inner side.

An intermediate side wall member is adapted to have its lower edge engage a respective said side groove in said first floor member, to have its front edge engage one of said vertical L-shaped front wall beams, and to have its top edge engage said downwardly disposed intermediate slotted beam in said third floor member, comprising an outwardly disposed vertical grooved beam member with an intermediate horizontal slot forming a notch therein, an intermediate horizontal slot extending forwardly from the rear edge, at least two holes disposed at the top and bottom rear corners, and open doorways cut into said wall member.

An extension front wall has a size and shape to have its bottom edge engage said groove that is parallel to

4

but not overlapping said front groove in said first floor member, to have its top edge engage said groove in said downwardly disposed front beam in said third floor member, and to have a side edge engage said groove in said vertical beam in said intermediate side wall member, and which comprises a vertical L-shaped beam member at one side edge, and an intermediate horizontal inwardly disposed grooved beam member.

An extension side wall has a size and shape to have its bottom edge engage said groove that is parallel to and shorter than said side grooves in said first floor member, to have its frontmost side edge engage said L-shaped beam in said extension front wall member, and to have its top edge engage said groove in said downwardly disposed side beam in said third floor member, and comprises an intermediate horizontal inwardly disposed grooved beam member and holes disposed at the rear of the top and bottom edges and in said grooved beam member.

A second floor member adapts to have its side and front edges engage said grooves of said intermediate inner beam members of said side, extension side, front, and extension front walks, and to engage said horizontal notch formed in said vertical intermediate beam in said intermediate side wall member, and comprising a stairwell near the front edge, at least two holes disposed at the far right and left edges near the rear edge, and at least two horizontal slots extending rearwardly from the front edge, said slots adapted to engage said horizontal slots of said intermediate and intermediate side wall members.

A side wall has a respective size and shape to have its front edge engage said L-shaped beam of said front wall member, and its bottom edge engage one of said side grooves as of said first floor member, comprising at least one inner horizontal beam member having an inward side groove, a top horizontal beam member containing a top inwardly slanted groove and an inward side groove, and at least two holes disposed at the lower rear corner and above it in an inner horizontal beam member.

At least one intermediate wall member is adapted to have its lower edge engage a respective said intermediate groove in said first floor and to have its forward edge engage said notches and vertical slots respectively in said segmented horizontal and top front beam members, comprising at least one intermediate horizontal slot extending forwardly from the rear edge; open doorways formed above and in extension of said intermediate horizontal slots, a recessed lower rear edge extending from the level of said first floor member to the level of the lowest of said intermediate horizontal slots, a forward edge having its upper portions angled inwardly, and holes disposed at the bottom, forward, and top edges.

A third floor member adapts to have its front edge and a side edge respectively engage said inward side grooves in said front and side wall top beam members, comprising a stairwell near the front edge, holes disposed along the front and side edges, a side horizontal beam disposed in a downward direction, a downwardly disposed intermediate grooved beam member parallel to and longer than said side beam, an upwardly disposed beam member directly above said downwardly disposed intermediate beam member with an inwardly slanting groove disposed therein, a downwardly disposed beam member perpendicular to and running between said side and intermediate beam members at the

5

foremost edge of said floor member, and at least one horizontal slot extending rearwardly from the front edge, said slot adapted to engage said horizontal slot of said intermediate wall member.

Front, left and right roof wall members are disposed with bottom edges in said slanted grooves in molding of said front and side top beam members and said intermediate upwardly disposed beam member in said third floor member, and said front roof member comprising left and right side L-shaped beam members, said front roof member having at least two dormer window units disposed therein.

A top roof member includes chimney holes disposed therein, left, right, and front beam members with outwardly slanted grooves disposed thereunder, a rear edge molding beam and said front beam member having inwardly disposed notches which correspond with said intermediate wall members and said chimney holes, said outwardly slanting grooves in said front, left, and right beams respectively engaging the top edges of said front, left and right roof members.

A chimney member contains a pair of parallel downwardly projecting members is adapted to be inserted in said roof hole, to straddle said center wall and to be secured thereto by a peg inserted in aligned holes in said center wall and said parallel members.

A plurality of pegs adapts to be inserted in holes contained in said beams, grooves, wall and floor members at their junctures with each other for retaining said assembled structure together.

An easily assembled knock-down type of doll house includes front and side wall members, front, side and top roof members, intermediate wall and floor members, and a doorway assembly including side door members with inner grooves disposed therein, step member, door members, door overhang member and cross member, wherein inner front wall edges are disposed in side inner grooves of said side door member whereby said doorway assembly is secured by and within the front wall member.

BRIEF DESCRIPTION OF THE DRAWINGS

FIG. 1 is a front perspective view of a doll house constructed according to the present invention.

FIG. 2 is a rear perspective view showing the doll house of FIG. 1.

FIG. 3 is an exploded view of the doll house of FIGS. 1 and 2.

FIG. 3A shows an L-shaped connector used between main and auxiliary front wall panels of the doll house.

FIG. 4 is a detail of a partially-assembled from of the doll house shown in FIGS. 1–3.

FIG. 5 is a detail of a main portion of the assembled skeletal structure of the doll house of the present invention.

FIG. 6 is a detail of the door of the present invention.

FIG. 7 is an exploded detail of the door assembly shown in FIG. 6.

FIG. 7A is a detail of the side of the door assembly and FIG. 7B is a detail of a side of the step portion of the door assembly.

DETAILED DESCRIPTION OF THE DRAWINGS

Referring to FIG. 1, the doll house is generally indicated by the numeral 1. The doll house has a base 2, a main front panel 4 with upper and lower window assemblies 2 and 6 respectively. A door assembly 10 is

6

fitted in the front panel 4. An auxiliary front panel 12 is provided and a side panel 14 is shown in FIG. 1.

Roof structure 20 includes a front roof panel 22 with gables 24, a top roof panel 26 and side roof panels 28. Chimneys 30 and railing 32 complete the elements shown in FIG. 1.

As shown in FIG. 2, the railing 32 has three separate elements 34, 36, and 38 which are connectable to the cantilevered deck portion 42 of floor member 40. The doll house is seen to have central relatively tall vertical members 50 and 52 and relatively short outer wall 54 and inner wall 56. Lower stair 60 is connectable to the base and upper stair 62 with railing 64 attached is connected between floor members 40 and 66. Upper railing 68 completes the structure.

As shown in FIG. 3, base member 2 has a series of connected grooves 70 which includes a long frontal groove 72 for receiving front wall 4.

Grooves 74 and 76 receive bottom edges of relatively long vertical wall members 50 and 52. Groove 78 receives the bottom edge of relatively short outer wall member 54 and groove 80 receives the lower edge of relatively short interior wall member 56. Groove 82 receives the lower edge of the second frontal wall member 12 and groove 84 receives the lower edge of side wall member 14.

As shown in the drawings the exposed edges of the vertical wall members 14, 50, 52, 54, and 56 are provided with vertically grooved strips 86 which provide a finished appearance and which abut the base at the end of grooves 84, 74, 76, 78, and 80 respectively to provide a finished appearance as well as providing additional support structure.

In FIG. 3 the elements are shown in a exploded view out of their normal position and some of the elements are shrunk while other of the elements are stretched or divided to emphasize the details. The arrangement and the relative size of the elements are best observed with relation to FIG. 2 and other assembled views.

As shown in FIG. 3, slotted moldings are provided an intersections of panel elements to provide secure interconnections. Molding element 90 at the top of front wall 4 has a rearward opening slot 92 to receive the forward edge of floor member 40. Vertical slots 94 in the molding receive forward edges of the relatively tall interior wall elements 50 and 52. Molding 96 has a slot 98 which receives the forward edge of intermediate floor member 100. Appropriate slits such as 102 in floor member 100 and 104 in vertical wall 52 cooperate to interengage the orthogonal related elements.

The connection between the front walls 4 and 12 and the interior wall 56 is schematically shown in detail on FIG. 3A. Element 106 has slot 108 for receiving a vertical edge of the front wall 4. Slot 110 receives a vertical inner edge of wall 12, and slot 112 receives the forward vertical edge of interior wall 56.

In initial steps of assembly a door and step assembly 10 are inserted in front wall 4. The front door assembly and the porch steps have grooves on both sides. The door assembly and porch steps are slid up into the door opening. When they are properly in place, the lower edge of the front wall extends slightly below the porch steps for fitting into groove 72 in the base. The lower edge of outer side wall 54 is fitted into groove 78 in the base and the forward edge of panel 54 fits within an angle strip glued along the flat edge of front panel 4.

Three pegs 114, 120 are then inserted in positions 122, 124, and 125 as shown in FIG. 4.

Bottom edges of the tall vertical wall members 50 and 52 are inserted in grooves 74 and 76 of base member 2 respectively, and pegs are inserted in positions 128 and 130. The bottom edge of vertical interior wall member 56 is inserted in slot 80 in base 2, and a peg is provided at position 132.

The third floor member 40 is then inserted through slots 134 and 138 of vertical members 50 and 52. The right edge 140 is carefully slid along slot 142 of member 144 which is glued at top wall 54. The front edge 146 of floor member 40 is carefully slid into slot 148 in the crown atop the front wall 4. The top edge of wall member 56 is inserted in a downwardly disposed intermediate grooved beam member in the third floor member 40. The structure then appears as shown in FIG. 5.

Pegs are inserted at positions 150, 152, and 156.

In the next steps, the cantilevered portion 42 of floor member 40 is slightly flexed upwardly. The second front panel 12 is inserted with its bottom edge 160 in groove 82 and its upper edge 162 in a downward opening groove in element 164 on floor member 40. Continuing to flex the cantilevered portion 42 upward, outer wall 14 has its lower edge 166 inserted in groove 84 of base 2 and its upper edge 168 inserted in a downward opening groove in end molding 170 of floor member 40. No pegs are required for the connections of walls 12 and 14. After the connection of the walls 12 and 14, the peg may be removed from position 128 without weakening the structure. The second floor member is then slid into place with related slots interfitting and edges 172, 174, and 178 fitting into appropriate grooves in moldings of the front and outer walls. Front roof member 22 is then placed in position with its lower edge 180 in an upward opening groove in molding 90 atop front wall 4. Upper edges of members 50 and 52 are fitted within grooves in roof top member 24 and a downward opening groove in molding 182 receives upper edge 184 of front roof member 22. Lugs 186 on chimneys 30 are fitted through openings 188 in the top roof member 24 and pins are inserted through openings 190 and complimentary openings 192 near the upper edges of vertical wall members 50 and 52.

Roof side panels 28 are slipped into place with lower edges 194 and 196 positioned in grooves 198 and 200 respectively in the wall member 54 and floor member 40. Upper edges 202 and 204 are inserted in downward opening grooves 206 and 208 in edge moldings of the upper roof member 24.

As shown in detail in FIGS. 6, 7, 7A, and 7B, door 10 has a door portion 210 and a step portion 212.

Door 210 has side members 214 and 216 which include side lights 218. Atop the side members 214 and 216 are mounted outer facial member 220 and interior facial member 222 as shown in detail in FIG. 7A. The side members have vertical grooves 224 which receive edges 226 of the door opening as shown in FIG. 3. Upper edge 228 of the door opening is received within groove 230 between the outer and inner facial elements 220 and 222. The lower edge 232 of the door assembly fits downward within upward groove 234 of the step assembly as shown in FIG. 7B. The step assembly 212 has vertical side grooves 236 which receives edges 226 of the door opening.

Because of the unique construction of the present invention, wall panel elements and rooms may be added to the doll house without adding connectors. The addition of such rooms enables reduction in connection elements.

8

The advantages of the invention may be obtained in multiplications such as those having a single vertical element and lateral extensions of the doll house in opposite directions.

Variations and modifications of the invention may be constructed without departing from the scope of the invention which is defined in the following claims.

What is claimed is:

1. An easily assembled knock-down doll house apparatus comprising a base having plural connected angularly oriented slots for receiving bottom edges of vertical walls, the slots comprising a first elongated slot slightly spaced inward from a first elongated edge of the base for receiving a front vertical wall member and plural slots communicating with the first elongated slot and extending angularly therefrom toward a second opposite elongated edge of the base, and a further slot extending angularly outward from an outer one of the plural slots and extending toward an edge of the base intermediate the first and second elongated edges, a remote slot extending from a portion of the further slot remote from the outer one of the plural slots toward the second elongated edge of the base, a front vertical wall member mounted in the first elongated slot, vertical transverse wall members mounted in the plural slots, the vertical transverse wall members comprising at least one relatively short outer wall member mounted in the outer one of the plural slots and at least one relatively tall inner wall member, a floor member fitted together with the vertical transverse wall members parallel to the base and spaced therefrom, the floor member extending over the relatively short outer wall member and through the relatively tall inner wall member, and the floor member cantilevering outward beyond the relatively short outer wall member and over the further slot and remote slot, a second front vertical wall member mounted in the further slot and extending between the base and a cantilevered portion of the floor member, a vertically extending connector element having a transverse portion extending between the edge of the front vertical wall member adjacent the outer one of the plural slots and the edge of the second front vertical wall member adjacent said outer one of the plural slots, a first longitudinal portion extending from an end of the transverse portion and overlying the front portion of the front vertical wall member adjacent the transverse portion, a second longitudinal portion extending from the opposite end of the transverse portion in a direction opposite the first longitudinal portion and overlying the rear portion of the second front vertical wall member adjacent the transverse portion, the portion of the transverse portion adjacent the second longitudinal portion being provided with a rearwardly opening vertical slot which receives the front vertical edge of the relatively short outer wall member, and means for connecting the base, the vertical transverse wall members and the floor member for forming thereby a rigid skeletal frame of the base, vertical transverse wall members and floor member by which a doll house comprising the members is rigidly supported.

2. The doll house in claim 1, further comprising a side vertical wall member mounted in the remote slot and extending between the base and a cantilevered portion of the floor member and having a first vertical edge fitted together with a vertical edge of the second front vertical wall member which is spaced from the relatively short outer wall member.

9

3. The apparatus of claim 1, wherein the relatively tall inner wall member comprises an inward and upward sloping frontal edge at positions thereof above the floor member and further comprising a roof front member having a lower edge fitted together with a front edge of the floor member, the roof front member sloping inwardly and resting against the inward sloped edge of the relatively tall inner wall member, a roof top member resting on the upper edge of the relatively tall inner wall member and having a front edge fitted together with an upper edge of the roof front member and means for connecting the roof top member to the relatively tall inner wall member thereby cooperating with the roof top member and the floor member to hold the roof front member in assembled condition, the floor member and the roof top member having complementary upward and downward opening slot means respectively, and first and second roof side members mounted in the slot means.

4. The doll house apparatus of claim 3, wherein upward opening slot means in the floor member are positioned immediately above the relatively short outer wall member and wherein a portion of the floor member cantilevered outward beyond the slot means comprises an outer deck member.

5. The apparatus of claim 1, wherein the relatively tall inner wall member is provided with a horizontal slit extending partially through the wall member and further comprising an intermediate floor member having a plan form similar to a plan form of the first said floor member and fitting within the slit and engaging in rearward opening slot means positioned on the front vertical wall member.

6. The apparatus of claim 1 wherein the front vertical wall member is provided with a door opening extending upward into the wall from a bottom edge of the wall, a door assembly for fitting in the opening, the door assembly comprising side members having outwardly facing vertical slots for receiving opposite vertical edges of the wall defining the door opening and a movable door hinged to one of the side members and abutting the other of said side members when in a closed position.

7. The apparatus of claim 6, wherein the door assembly further comprises a top member connected to the side members, the top member having an upward opening horizontal slot for receiving an upper edge of the wall defining the door opening.

8. The apparatus of claim 6, further comprising a step assembly having opposite outward opening vertical grooves for receiving vertical edges of the wall defining the door opening.

9. The easily assembled knock-down doll house apparatus of claim 1 wherein:

the base comprises a generally rectangular member having a series of in-line holes disposed behind said first elongated slot,

the front vertical wall member has at least one segmented horizontal grooved molding member disposed on the inner side, one top horizontal molding member containing a top inwardly slanted groove and a rearward opening horizontal slot and an inwardly disposed vertical slot aligned with the space between the segments of the segmented horizontal grooved molding member and the plural slot in said base receiving the relatively tall inner wall member, a cutout portion adjacent said first elongated slot and at least two holes disposed at the bottom edge of the front vertical wall member,

10

the relatively short outer wall member has its top edge engaging a downwardly disposed intermediate grooved beam member in said floor member, and further has an intermediate horizontal slot extending forwardly from the rear edge, at least one hole disposed at the bottom rear corner, and open doorways formed therein,

the second front vertical wall member has its top edge engaging a groove in a downwardly disposed front beam member in said floor member, and further has an intermediate horizontal inwardly disposed grooved beam member,

a side vertical wall member is mounted in the remote slot, said side vertical wall member having its frontmost side edge fitted together with said second front vertical wall member, and having its top edge engage a groove in a downwardly disposed side horizontal beam in said floor member, and further having an intermediate horizontal inwardly disposed grooved beam member,

the relatively tall inner wall member has its forward edge engaging said spaces and vertical slot respectively in said segmented horizontal grooved molding member and top horizontal molding member, and further has plural vertically spaced intermediate horizontal slots extending forwardly from the rear edge; open doorways formed above and in extension of said intermediate horizontal slots, a recessed lower rear edge extending from the level of said base to the level of the lowest of said intermediate horizontal slots, a forward edge having its upper portions angled inwardly, and holes disposed at the bottom forward, and top edges,

a second floor member is provided having its side edge engage said groove in said intermediate horizontal inwardly disposed grooved beam member of said side vertical wall member and its front edge engage said groove in said segmented horizontal molding member of said front vertical wall member, and further having a stairwell near the front edge, and at least two horizontal slots extending rearwardly from the front edge, said slots adapted to engage said intermediate horizontal slots of said relatively short outer wall member and said relatively tall inner wall members,

a second side vertical wall member is provided having a respective size and shape to have its front edge fitted together with said front vertical wall member, and its bottom engage one of said plural slots in said base, comprising at least one inner horizontal beam member having an inwardly opening side groove, a top horizontal beam member containing a top inwardly slanted groove and an inwardly opening side groove and at least one hole disposed at the lower rear corner,

the floor member is adapted to have its front edge and a side edge respectively engage said rearward opening horizontal slot in said top horizontal molding member of said front vertical wall member and the inwardly opening side groove in the top horizontal beam member of the second side vertical wall member, said floor member further comprising a stairwell near the front edge, holes disposed along the front and side edges, a side horizontal beam disposed in a downward direction, a downwardly disposed intermediate grooved beam member parallel to and longer than said side horizontal beam, an upwardly disposed top horizontal beam

Patentsammlung Puppenhäuser I • Design und Technik • Reihe: Vorlage – Konzept – Konstruktion • Band: 1
Herausgeber: www.atelier-kalai.de, Kerstin Winter • Hersteller / Verlag: Books on Demand GmbH Norderstedt